CliffsNotes® Anatomy & Physiology Quick Review

By Phillip E. Pack, Ph.D., and Steven Bassett

2nd Edition

Houghton Mifflin Harcourt
Boston   New York

About the Authors

Phillip E. Pack has taught AP courses and gifted programs for 11 years. He is currently an assistant professor of Math and Science at Woodbury University in Burbank, California.

Steven Bassett has taught Anatomy and Physiology courses to undergraduates for over 21 years and Pathophysiology to physician assistants for 10 years. He has been at Southeast Community College in Lincoln, Nebraska since 1990.

Authors' Acknowledgements

The authors would like to thank Grace Freedson for bringing us this project. We also want to thank our families for their love and support.

Publisher's Acknowledgments

Editorial

Acquisitions Editor: Greg Tubach
Project Editor: Suzanne Snyder
Copy Editor: Lynn Northrup
Technical Editors: Robin Vance,
    Colonel (ret.) Michael Yard

Composition

Indexer: BIM Indexing & Proofreading Services
Proofreader: Laura Bowman
Wiley Publishing, Inc. Composition Services

CliffsNotes® Anatomy & Physiology Quick Review, 2nd Edition

Library of Congress Control Number: 2011930127
ISBN: 978-0-470-87874-3 (pbk), 978-1-118-08457-1 (ebk), 978-1-118-08458-8 (ebk), 978-1-118-08459-5 (ebk)
Printed in the United States of America
DOC 10 9 8 7   4500574277

For information about permission to reproduce selections from this book, please write Permissions, Houghton Mifflin Harcourt Publishing Company, 215 Park Avenue South, New York, New York 10003.

www.hmhco.com

Table of Contents

INTRODUCTION

Everyone, from high school students to medical students, needs to have a basic knowledge of human anatomy and physiology. If you understand how your body is built and the different functions it performs, you will likely appreciate it more than you probably do.

The human body is complex and houses many systems. A general grasp of biology is helpful in understanding anatomy, but not necessary, while a general knowledge of chemistry is beneficial in comprehending physiology. Don't worry if you don't have that knowledge; this book gives you the basics so you can understand the rest.

Why You Need This Book

Can you answer yes to any of these questions?

- Do you need to review the fundamentals of anatomy and physiology fast?
- Do you need a course supplement to human anatomy and physiology?
- Do you need a concise, comprehensive reference for anatomy and physiology?

If so, then *CliffsNotes Anatomy & Physiology Quick Review,* 2nd Edition, is for you!

How to Use This Book

You're in charge here. You get to decide how to use this book. You can read it straight through or just look for the information that you want and then put the book back on the shelf for later use. Here are a few of the recommended ways to search for information about a particular topic:

- Look for areas of interest in the book's table of contents or use the index to find specific topics.
- Flip through the book, looking for subject areas at the top of each page.

■ Get a glimpse of what you'll gain from a chapter by reading through the "Chapter Check-In" at the beginning of each chapter.

■ Use the "Chapter Check-Out" at the end of each chapter to gauge your grasp of the important information you need to know.

■ Test your knowledge more completely in the Review Questions and find additional sources of information in the Resource Center.

■ Look in the glossary for important terms and definitions. If a word is boldfaced in the text, you can find a more complete definition in the glossary.

Hundreds of Practice Questions Online!

Go to www.cliffsnotes.com/sciences/anatomy-and-physiology-quizzes for hundreds of additional anatomy and physiology practice questions to help prepare you for your next quiz or test. The questions are organized by this book's chapter sections, so it's easy to use the book and then quiz yourself online to make sure you know the subject. Go to www.cliffsnotes.com to test yourself anytime and find other free homework help.

Chapter 1

ANATOMY AND CHEMISTRY BASICS

Chapter Check-In

❑ Understanding the basics of anatomy

❑ Noting the basic chemical constituents that help form matter

❑ Listing the types of bonds that form between two atoms

❑ Understanding the difference between inorganic and organic compounds

❑ Describing the four classes of organic molecules

❑ Finding out how a chemical reaction occurs in a biological system

After you know the basic terms of anatomy but before studying the structure and function of the body, you need to have a basic knowledge of chemistry that will be pertinent to your studies. Some of the chemistry presented in this chapter may not be new to you. In fact, the organic molecules of carbohydrates, lipids (such as fats, cholesterol, and steroids), and proteins are the staples of a healthy diet and lifestyle. Learning these basic chemical components is essential for future studies in physiology, nutrition, and many other fields of scientific interest.

What Is Anatomy and Physiology?

Anatomy is the study of the structure and relationship between body parts. **Physiology** is the study of the function of body parts and the body as a whole. Some specializations within each of these sciences follow:

- *Gross (macroscopic) anatomy* is the study of body parts visible to the naked eye, such as the heart or bones.

- *Histology* is the study of tissues at the microscopic level.

- *Cytology* is the study of cells at the microscopic level.

- *Neurophysiology* is the study of how the nervous system functions.

Organizations of living systems

Living systems can be defined from various perspectives, from the broad (looking at the entire earth) to the minute (individual atoms). Each perspective provides information about how or why a living system functions:

- At the chemical level, ***atoms, molecules*** (combinations of atoms), and the chemical bonds between atoms provide the framework upon which all living activity is based.

- The ***cell*** is the smallest unit of life. **Organelles** within the cell are specialized bodies performing specific cellular functions. Cells themselves may be specialized. Thus, there are nerve cells, bone cells, and muscle cells.

- A *tissue* is a group of similar cells performing a common function. Muscle tissue, for example, consists of muscle cells.

- An *organ* is a group of different kinds of tissues working together to perform a particular activity. The heart is an organ composed of muscle, nervous, connective, and epithelial tissues.

- An *organ system* is two or more organs working together to accomplish a particular task. The digestive system, for example, involves the coordinated activities of many organs, including the mouth, stomach, small and large intestines, pancreas, and liver.

- An *organism* is a system possessing the characteristics of living things—the ability to obtain and process energy, the ability to respond to environmental changes, and the ability to reproduce.

Homeostasis

A characteristic of all living systems is **homeostasis,** or the maintenance of stable, internal conditions within specific limits. In many cases, stable conditions are maintained by negative feedback.

In **negative feedback,** a sensing mechanism (a receptor) detects a change in conditions beyond specific limits. A control center, or integrator (often the brain), evaluates the change and activates a second mechanism (an **effector**) to correct the condition; for example, cells that either remove or add glucose to the blood in an effort to maintain homeostasis are effectors. Conditions are constantly monitored by receptors and evaluated by the control center. When the control center determines that conditions have returned to normal, corrective action is discontinued. Thus, in negative feedback, the variant condition is canceled, or negated, so that conditions are returned to normal.

The regulation of glucose concentration in the blood illustrates how homeostasis is maintained by negative feedback. After a meal, the absorption of glucose (a sugar) from the digestive tract increases the amount of glucose in the blood. In response, specialized cells in the pancreas (alpha cells) secrete the hormone insulin, which circulates through the blood and stimulates liver and muscle cells to absorb the glucose. Once blood glucose levels return to normal, insulin secretion stops. Later, perhaps after heavy exercise, blood glucose levels may drop because muscle cells absorb glucose from the blood and use it as a source of energy for muscle contraction. In response to falling blood glucose levels, another group of specialized pancreatic cells (beta cells) secretes a second hormone, glucagon. Glucagon stimulates the liver to release its stored glucose into the blood. When blood glucose levels return to normal, glucagon secretion stops.

Compare this with **positive feedback,** in which an action intensifies a condition so that it is driven farther beyond normal limits. Such positive feedback is uncommon but does occur during blood clotting, childbirth (labor contractions), lactation (where milk production increases in response to an increase in nursing), and sexual orgasm.

Anatomic terminology

In order to accurately identify areas of the body, clearly defined anatomical terms are used. These terms refer to the body in the anatomical position—standing erect, facing forward, arms down at the side, with the palms turned forward. In this position, the following apply:

■ Directional terms are used to describe the relative position of one body part to another. These terms are listed in Table 1-1.

■ Body planes and sections are used to describe how the body or an organ is divided into two parts:

  ■ *Sagittal planes* divide a body or organ vertically into right and left parts. If the right and left parts are equal, the plane is a midsagittal plane; if they're unequal, the plane is a parasagittal plane.

  ■ A *frontal (coronal) plane* divides the body or organ vertically into front (anterior) and rear (posterior) parts.

  ■ A *horizontal (transverse) plane* divides the body or organ horizontally into top (superior) and bottom (inferior) parts. This is also known as a cross-section.

■ Body cavities are enclosed areas that house organs. These cavities are organized into two groups:

  ■ The *posterior/dorsal* body cavity includes the cranial cavity (which contains the brain) and the vertebral cavity (which contains the spinal cord).

  ■ The *anterior/ventral* body cavity includes the thoracic cavity (which contains the lungs, each in its own pleural cavity, and the heart, in the pericardial cavity) and the abdominopelvic cavity (which contains the digestive organs in the abdominal cavity and the bladder and reproductive organs in the pelvic cavity).

■ Regional terms identify specific areas of the body. In some cases, a descriptive word is used to identify the location. For example, the axial region refers to the main axis of the body—the head, neck, and trunk. The appendicular region refers to the appendages—the arms and legs. Other regional terms use a body part to identify a particular region of the body. For example, the nasal region refers to the nose.

Table 1-1   Basic Anatomy Terms

Term	Definition	Example
Superior	Above another structure.	The heart is superior to the stomach.
Inferior	Below another structure.	The stomach is inferior to the heart.
Anterior/ventral	Toward the front of the body.	The navel is anterior to the spine.
Posterior/dorsal	Toward the back of the body.	The spine is posterior to the navel.
Medial	Toward the midline of the body. (The midline divides the body into equal right and left sides.)	The nose is medial to the eyes.
Lateral	Away from the midline of the body (or toward the side of the body).	The ears are lateral to the nose.
Ipsilateral	On the same side of the body.	The spleen and descending colon are ipsilateral.
Contralateral	On opposite sides of the body.	The ascending and descending portions of the colon are contralateral.
Intermediate	Between two structures.	The knee is intermediate between the upper leg and lower leg.
Proximal	Closer to the point of attachment of a limb.	The elbow is proximal to the wrist.
Distal	Farther from the point of attachment of a limb.	The foot is distal to the knee.
Superficial	Toward the surface of the body.	The skin is superficial to the muscle.
Deep	Away from the surface of the body.	The skeleton is deep to the skin.

Atoms, Molecules, Ions, and Bonds

Matter is anything that takes up space and has mass. Matter consists of elements that possess unique physical and chemical properties. Elements are represented by chemical symbols of one or two letters, such as C (carbon), Ca (calcium), H (hydrogen), O (oxygen), N (nitrogen), and P (phosphorus). The smallest quantity of an element that still possesses the characteristics of that element is an **atom.** Atoms chemically bond together to form *molecules,* and the composition of a molecule is given by its chemical formula (O_2, H_2O, $C_6H_{12}O_6$). When the atoms in a molecule are different, the molecule is a *compound* (H_2O and $C_6H_{12}O_6$, but not O_2).

The atoms of the elements consist of a nucleus containing positively charged *protons* and neutrally charged *neutrons*. Negatively charged *electrons* are arranged outside the nucleus. The atoms of each element differ by their number of protons, neutrons, and electrons. For example, hydrogen has one proton, one electron, and no neutrons, while carbon has six protons, six neutrons, and six electrons. The number and arrangement of electrons of an atom determine the kinds of chemical bonds that it forms and how it reacts with other atoms to form molecules. There are three kinds of chemical bonds:

- *Ionic bonds* form between two atoms when one or more electrons are completely transferred from one atom to the other. The atom that gains electrons has an overall negative charge, and the atom that donates electrons has an overall positive charge. Because of their positive or negative charge, these atoms are *ions.* The attraction of the positive ion to the negative ion constitutes the ionic bond. Sodium (Na) and chlorine (Cl) form ions (Na^+ and Cl^-), which attract one another to form the ionic bond in a sodium chloride (NaCl) molecule. A plus or minus sign following a chemical symbol indicates an ion with a positive or negative charge, which results from the loss or gain of one or more electrons, respectively. Numbers preceding the charges indicate ions whose charges are greater than one (Ca^{2+}, PO_4^{3-}).

- *Covalent bonds* form when electrons are shared between atoms. That is, neither atom completely retains possession of the electrons (as happens with atoms that form ionic bonds). A single covalent bond occurs when two electrons are shared (one from each atom). A double or triple covalent bond is formed when four or six electrons are shared, respectively. When the two atoms sharing electrons are exactly the same, as in a molecule of oxygen gas (two oxygen atoms to form O_2), the electrons are shared equally, and the bond is a nonpolar covalent bond. When the atoms are different, such as in a molecule of water (H_2O), the larger nucleus of the oxygen atom exerts a stronger pull on the shared electrons than does the single proton that makes up either hydrogen nucleus. In this case, a polar covalent bond is formed because the unequal distribution of the electrons creates areas within the molecule that have either a negative or positive charge (or pole), as shown in Figure 1-1.

- *Hydrogen bonds* are weak bonds that form between the partially positively charged hydrogen atom in one covalently bonded molecule and the partially negatively charged area of another covalently bonded molecule. An individual water molecule develops a partially positively charged end and a partially negatively charged end; see Figure 1-1(a). Hydrogen bonds form between adjacent water molecules. Since the atoms in water form a polar covalent bond, the positive area in H_2O around the hydrogen proton attracts the negative areas in an adjacent H_2O molecule. This attraction forms the hydrogen bond; see Figure 1-1(b).

Figure 1-1 Two examples of chemical bonds.

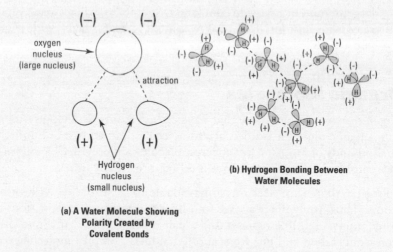

(a) A Water Molecule Showing
Polarity Created by
Covalent Bonds

(b) Hydrogen Bonding Between
Water Molecules

Inorganic Compounds

Inorganic compounds are typically compounds without carbon atoms. H_2O, O_2, and NaCl are examples of inorganic compounds.

Water is the most abundant substance in the body. Its abundance is due partly to its unique chemical properties created by the influence of its hydrogen bonds. These properties include the following:

- *Solvency.* Water is an excellent solvent. Ionic substances are soluble in water (they dissolve) because the poles of the polar water molecules pull them apart, forming ions. Polar covalent substances are also water-soluble because they share the same hydrogen bonding as water shares with itself. For this reason, polar covalent substances are called **hydrophilic** (water loving). Because they lack charged poles, nonpolar covalent substances do not dissolve in water and are called **hydrophobic** (water fearing).

- *Cohesion.* Because water molecules are held together by hydrogen bonds, water molecules have a high degree of cohesion, or the ability to stick together. As a result, water has strong surface tension. This tension, in turn, gives water strong capillary action, allowing water to creep up narrow tubing. These qualities contribute to the movement of water through capillaries.

- *Stability.* The temperature of water is stable. You must add a relatively large amount of energy to warm (and boil) it and remove a large amount of energy to cool (and freeze) it. So, when sweat evaporates from your forehead, a large amount of heat is taken with it and you are cooled.

Organic Molecules

Organic compounds are those that have carbon atoms. In living systems, large organic molecules, called macromolecules, can consist of hundreds or thousands of atoms. Most macromolecules are polymers, molecules that consist of a single unit (monomer) repeated many times.

Four of carbon's six electrons are available to form bonds with other atoms. Thus, you will always see four lines connecting a carbon atom to other atoms, each line representing a pair of shared electrons (one electron from carbon and one from another atom). Complex molecules can be formed by stringing carbon atoms together in a straight line or by connecting carbons together to form rings. The presence of nitrogen, oxygen, and other atoms adds variety to these carbon molecules.

Four important classes of organic molecules—carbohydrates, lipids, proteins, and nucleic acids—are discussed in the following sections.

Carbohydrates

Carbohydrates are classified into three groups according to the number of sugar (or saccharide) molecules present:

■ A *monosaccharide* is the simplest kind of carbohydrate. It is a single sugar molecule, such as a fructose or glucose (Figure 1-2). Sugar molecules have the formula $(CH_2O)_n$, where n is any number from 3 to 8. For glucose, n is 6, and its formula is $C_6H_{12}O_6$. The formula for fructose is also $C_6H_{12}O_6$, but as you can see in Figure 1-2, the placement of the carbon atoms is different. Very small changes in the position of certain atoms, such as those that distinguish glucose and fructose, can dramatically change the chemistry of a molecule.

■ A *disaccharide* consists of two linked sugar molecules. Glucose and fructose, for example, link to form sucrose (see Figure 1-2).

■ A *polysaccharide* consists of a series of connected monosaccharides. Thus, a polysaccharide is a polymer because it consists of repeating units of monosaccharide. Starch is a polysaccharide made up of a thousand or more glucose molecules and is used in plants for energy storage. A similar polysaccharide, glycogen, is used in animals for the same purpose.

Figure 1-2   The molecular structure of several carbohydrates.

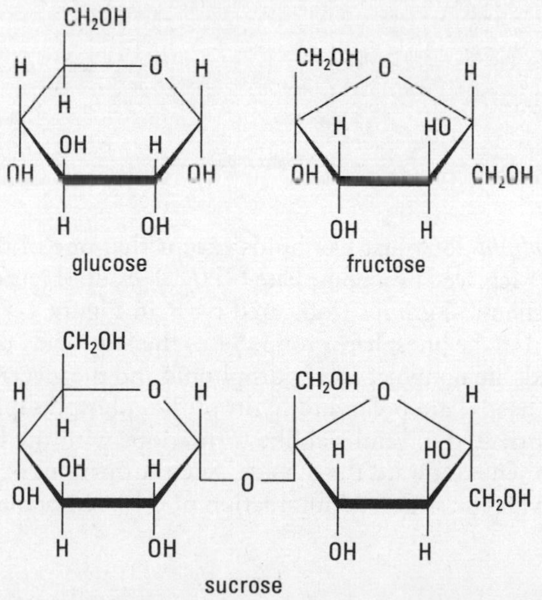

glucose

fructose

sucrose

Lipids

Lipids are a class of substances that are insoluble in water (and other polar solvents), but are soluble in nonpolar substances (such as ether or chloroform). There are three major groups of lipids:

■ *Triglycerides* include fats, oils, and waxes. They consist of three fatty acids bonded to a glycerol molecule (Figure 1-3). Fatty acids are hydrocarbons (chains of covalently bonded carbons and hydrogens) with a carboxyl group (–COOH) at one end of the chain. A saturated fatty acid has a single covalent bond between each pair of carbon atoms, and each carbon has two hydrogens bonded to it. You can remember this fact by thinking that each carbon is "saturated" with hydrogen. An unsaturated fatty acid occurs when a double covalent bond replaces a single covalent bond and two hydrogen atoms (Figure 1-3). Polyunsaturated fatty acids have many of these double bonds.

Figure 1-3 The molecular structure of a triglyceride.

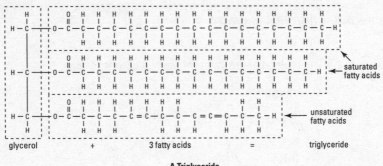

A Triglyceride

■ *Phospholipids* look just like lipids except that one of the fatty acid chains is replaced by a phosphate ($-PO_4^{3-}$) group (Figure 1-4). Additional chemical groups (indicated by R in Figure 1-4) are usually attached to the phosphate group. Since the fatty acid "tails" of phospholipids are nonpolar and hydrophobic and the glycerol and phosphate "heads" are polar and hydrophilic, phospholipids are often found oriented in sandwichlike formations with the hydrophobic heads oriented toward the outside. Such formations of phospholipids provide the structural foundation of cell membranes.

Figure 1-4 The molecular structure of a phospholipid.

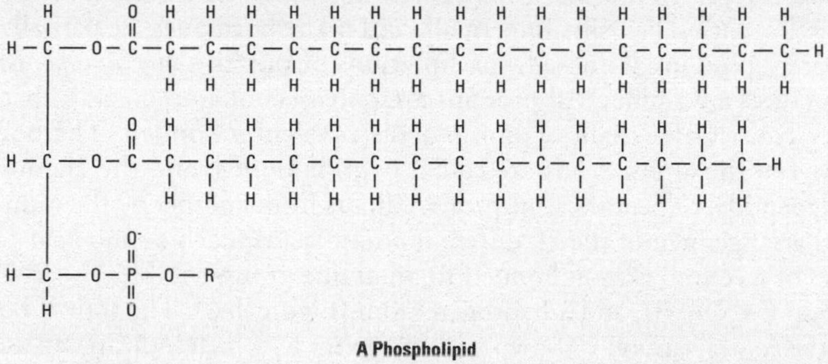

A Phospholipid

- **Steroids** are characterized by a backbone of four linked carbon rings (Figure 1-5). Examples of steroids include cholesterol (a component of cell membranes) and certain hormones, including testosterone and estrogen.

Figure 1-5 Examples of steroids.

steroid backbone

testosterone

cholesterol

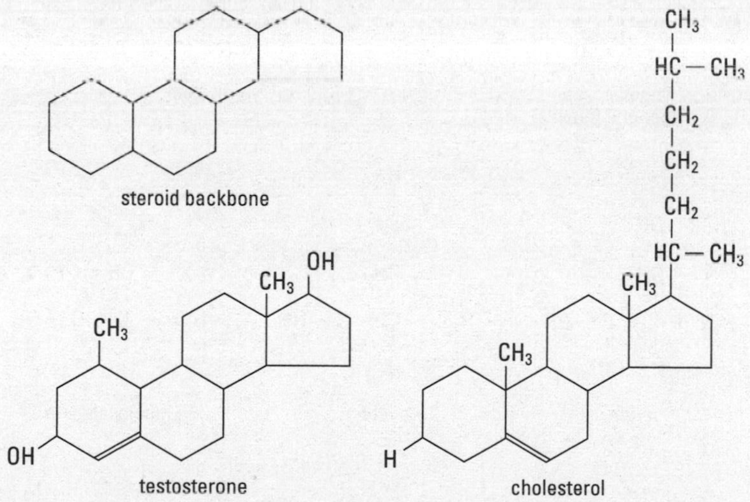

Proteins

Proteins represent a class of molecules that have varied functions. Eggs, muscles, antibodies, silk, fingernails, and many hormones are partially or entirely proteins. Although the functions of proteins are diverse, their structures are similar. All proteins are polymers of amino acids; that is, they consist of a chain of amino acids covalently bonded. The bonds between the amino acids are called peptide bonds, and the chain is a polypeptide, or peptide. One protein differs from another by the number and arrangement of the 20 different amino acids. Each amino acid consists of a central carbon bonded to an amine group ($-NH_2$), a carboxyl group ($-COOH$), and a hydrogen atom (Figure 1-6). The fourth bond of the central carbon is shown with the letter R, which indicates an atom or group of atoms that varies from one kind of amino acid to another. For the simplest amino acid, glycine, the R is a hydrogen atom. For serine, R is CH_2OH. For other amino acids, R may contain sulfur (as in cysteine) or a carbon ring (as in phenylalanine).

Figure 1-6 Examples of amino acids.

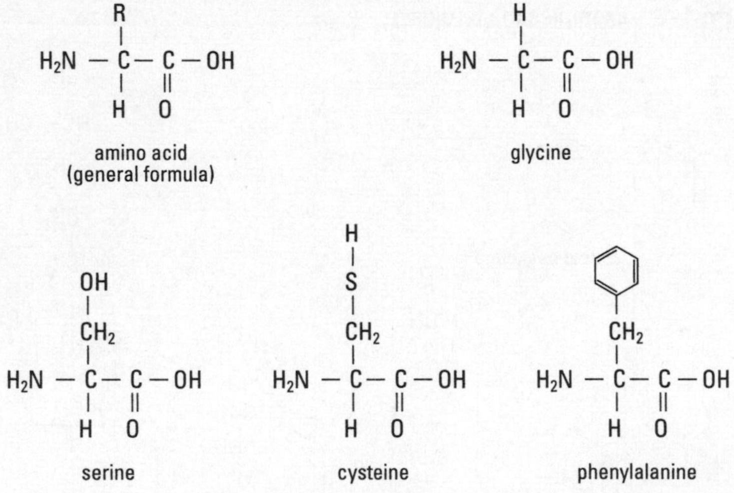

amino acid
(general formula)

glycine

serine

cysteine

phenylalanine

There are four levels that describe the structure of a protein:

- The primary structure of a protein describes the order of amino acids. Using three letters to represent each amino acid, the primary structure for the protein antidiuretic hormone (ADH) can be written as cys-tyr-glu-asn-cys-pro-arg-gly.

- The secondary structure of a protein is a three-dimensional shape that results from hydrogen bonding between amino acids. The bonding produces a spiral (alpha helix) or a folded plane that looks much like the pleats on a skirt (beta pleated sheet).

- The tertiary structure of a protein includes additional three-dimensional shaping that results from interaction among R groups. For example, hydrophobic R groups tend to clump toward the inside of the protein, while hydrophilic R groups clump toward the outside of the protein. Additional three-dimensional shaping occurs when the amino acid cysteine bonds to another cysteine across a disulfide bond. This causes the protein to twist around the bond (Figure 1-7).

Figure 1-7   Disulfide bonds can dictate a protein's structure.

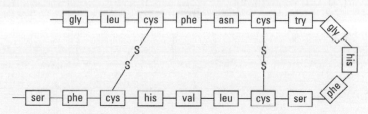

- The quaternary structure describes a protein that is assembled from two or more separate peptide chains. The protein hemoglobin, for example, consists of four peptide chains that are held together by hydrogen bonding, interactions among R groups, and disulfide bonds.

Nucleic acids

The genetic information of a cell is stored in molecules of **deoxyribonucleic acid (DNA)**. The DNA, in turn, passes its genetic instructions to **ribonucleic acid (RNA)** for directing various metabolic activities of the cell.

DNA is a polymer of nucleotides (Figure 1-8). A DNA molecule consists of three parts—a nitrogenous base, a five-carbon sugar called deoxyribose, and a phosphate group. There are four DNA nucleotides, each with one of the four nitrogenous bases (adenine, thymine, cytosine, and guanine). The first letter of each of these four bases is often used to symbolize the respective nucleotide (A for adenine nucleotide, for example).

Figure 1-8 The molecular structure of nucleotides.

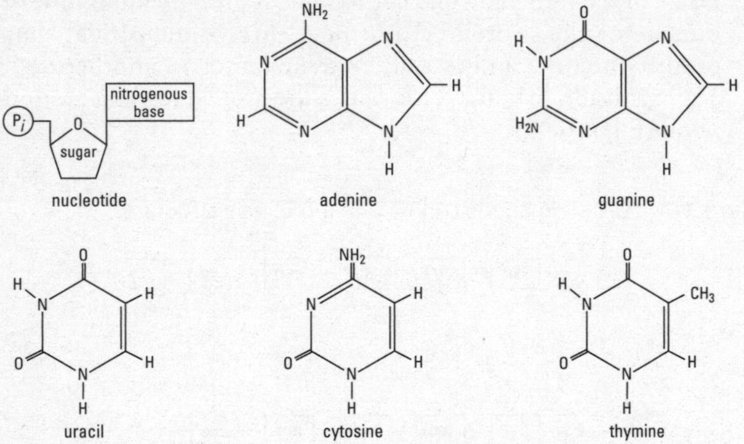

Figure 1-9 shows how two strands of nucleotides, paired by weak hydrogen bonds between the bases, form a double-stranded DNA. When bonded in this way, DNA forms a two-stranded spiral, or double helix. Note that adenine always bonds with thymine and cytosine always bonds with guanine.

RNA differs from DNA in the following ways:

- The sugar in the nucleotides that make an RNA molecule is ribose, not deoxyribose as it is in DNA.

- The thymine nucleotide does not occur in RNA. It is replaced by uracil. When pairing of bases occurs in RNA, uracil (instead of thymine) pairs with adenine.

- RNA is usually single-stranded and does not form a double helix as does DNA.

Figure 1-9   Two-dimensional illustrations of the structure of DNA.

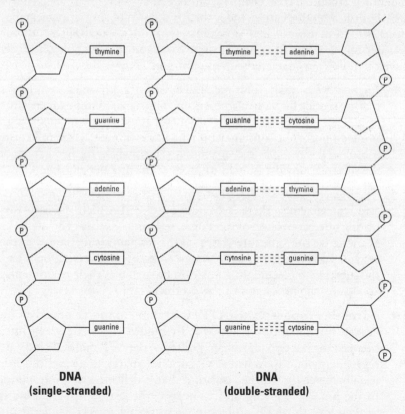

DNA
(single-stranded)

DNA
(double-stranded)

Chemical Reactions in Metabolic Processes

In order for a chemical reaction to take place, the reacting molecules (or atoms) must first collide and then have sufficient energy (activation energy) to trigger the formation of new bonds. Although many reactions can occur spontaneously, the presence of a catalyst accelerates the rate of the reaction because it lowers the activation energy required for the reaction to take place. A **catalyst** is any substance that accelerates a reaction but does not undergo a chemical change itself. Since the catalyst is not changed by the reaction, it can be used over and over again.

Chemical reactions that occur in biological systems are referred to as metabolism. **Metabolism** includes the breakdown of substances (catabolism), the formation of new products (synthesis or anabolism), or the transferring of energy from one substance to another. Metabolic processes have the following characteristics in common:

- *Enzymes* act as catalysts for metabolic reactions. Enzymes are proteins that are specific for particular reactions. The standard suffix for enzymes is "ase," so it is easy to identify enzymes that use this ending (although some do not). The substance on which the enzyme acts is called the substrate. For example, the enzyme amylase catalyzes the breakdown of the substrate amylose (starch) to produce the product glucose.

  The *induced-fit model* describes how enzymes work. Within the protein (the enzyme), there is an active site with which the reactants readily interact because of the shape, polarity, or other characteristics of the active site. The interaction of the reactants (substrate) and the enzyme causes the enzyme to change shape. The new position places the substrate molecules in a position favorable to their reaction and accelerates the formation of the product.

- **Adenosine triphosphate (ATP)** is a common source of activation energy for metabolic reactions. In Figure 1-10, the wavy lines between the last two phosphate groups of the ATP molecule indicate high-energy bonds. When ATP supplies energy to a reaction, it is usually the energy in the last bond that is delivered to the reaction. In the process of giving up this energy, the last phosphate bond is broken and the ATP molecule is converted to ADP (adenosine diphosphate) and a phosphate group (indicated by P_i). In contrast, new ATP molecules are assembled by phosphorylation when ADP combines with a phosphate group using energy obtained from some energy-rich molecule (like glucose).

Figure 1-10   The high-energy bonds of adenosine triphosphate (ATP).

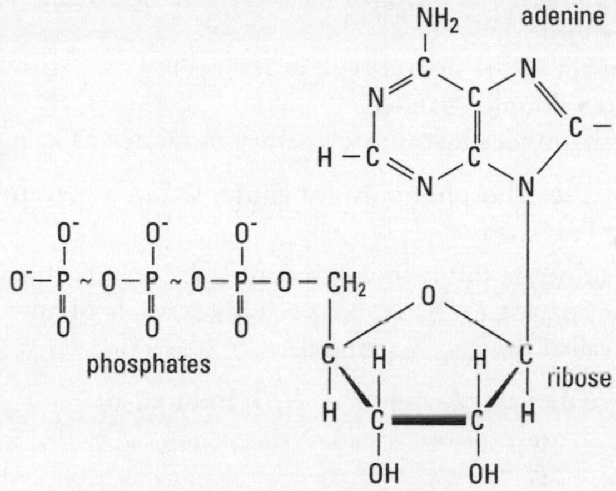

Adenosine Triphosphate (ATP)

■ **Cofactors** are nonprotein molecules that assist enzymes. A holoenzyme is the union of the cofactor and the enzyme (called an apoenzyme when part of a holoenzyme). If cofactors are organic, they are called **coenzymes** and usually function to donate or accept some component of a reaction, often electrons. Some vitamins are coenzymes or components of coenzymes. Inorganic cofactors are often metal ions, such as Fe^{++}.

Chapter Check-Out

Q&A

1. In metabolism, the breakdown of substances is called _____.

2. Which of the following protein structures describes only the amino acid sequence rather than its shape?

   a.   Primary structure
   b.   Secondary structure
   c.   Tertiary structure

3. Which of the following is true of RNA?

    a. Is composed of a nitrogen base, a six-carbon sugar, and a phosphate group

    b. Does not utilize deoxyribose as its sugar

    c. Is often double-stranded

    d. Has thymine, adenosine, cytosine, and uracil as its nucleotides

4. True or False: Phospholipids are composed of a glycerol molecule and three fatty acids.

5. When a substrate binds to an enzyme's active site, this interaction causes the enzyme to change shape. This example of how an enzyme works is called the _____ model.

Answers: 1. catabolism, **2.** a, **3.** b, **4.** F, **5.** induced-fit

Chapter 2
THE CELL

Chapter Check-In

❑ Discovering the functions and constituents of the plasma membrane

❑ Understanding the metabolic activities of the various organelles of a cell

❑ Identifying the mechanisms by which cells communicate and acquire vital substances

❑ Finding out the differences between mitosis and meiosis

❑ Detailing the process of synthesizing proteins via transcription, RNA processing, and translation

The cell is the smallest functional unit on which all life is built. Therefore, a strong knowledge of the various cellular organelles and their functions is crucial to any physiologist or anatomist. Identifying the components of each cell will help you ascertain not only the type of cell you're viewing, but also its function.

In order to perform the many diverse metabolic activities in the body, cells need to be able to communicate with one another to regulate growth and development. This is accomplished via genetic instructions, or DNA, contained within each cell. The processes by which this genetic information is replicated and utilized to build proteins are discussed throughout this chapter.

The Cell and Its Membrane

The **cell** is the basic functional unit of all living things. The plasma membrane (cell membrane) bounds the cell and encloses the nucleus (discussed presently) and *cytoplasm*. The cytoplasm consists of specialized bodies called organelles suspended in a fluid matrix, the cytosol, which consists of water and dissolved substances such as proteins and nutrients.

The plasma membrane

The *plasma membrane* separates internal metabolic events from the external environment and controls the movement of materials into and out of the cell. The plasma membrane is a double phospholipid membrane (lipid bilayer), with the nonpolar hydrophobic tails pointing toward the inside of the membrane and the polar hydrophilic heads forming the inner and outer faces of the membrane (Figure 2-1).

Proteins and cholesterol molecules are scattered throughout the flexible phospholipid membrane. Proteins may attach loosely to the inner or outer surface of the plasma membrane (peripheral proteins), or they may lie across the membrane, extending from inside to outside (integral proteins). The mosaic nature of scattered proteins within a flexible matrix of phospholipid molecules describes the fluid mosaic model of the cell membrane. Additional features of the plasma membrane follow:

- The phospholipid bilayer is semi-permeable. Only small, uncharged, polar molecules, such as H_2O and CO_2, and hydrophobic molecules—nonpolar molecules such as O_2 and lipid soluble molecules such as hydrocarbons—can freely cross the membrane.

- *Channel proteins* provide passageways through the membrane for certain hydrophilic (water-soluble) substances such as polar and charged molecules.

- *Transport proteins* spend energy (ATP) to transfer materials across the membrane. When energy is used to provide a passageway for materials, the process is called *active transport.*

- *Recognition proteins* (glycoproteins) distinguish the identity of neighboring cells. These proteins have oligosaccharide (short polysaccharide) chains extending from their cell surface.

- *Adhesion proteins* attach cells to neighboring cells or provide anchors for the internal filaments and tubules that give stability to the cell.

Figure 2-1 The phospholipid bilayer of the plasma membrane.

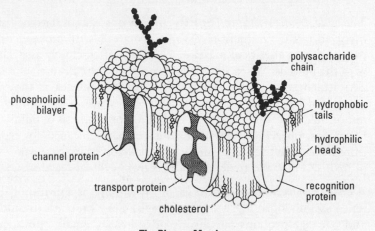

The Plasma Membrane

- *Receptor proteins* initiate specific cell responses once hormones or other trigger molecules bind to them.

- *Electron transfer proteins* are involved in moving electrons from one molecule to another during chemical reactions.

The nucleus and other organelles

Organelles are bodies within the cytoplasm that serve to physically separate the various metabolic activities that occur within cells. They include the following (Figure 2-2):

- The *nucleus* is bounded by the nuclear envelope, a phospholipid bilayer similar to the plasma membrane. The nucleus contains DNA (deoxyribonucleic acid), the hereditary information of the cell. Normally, the DNA is spread out within the nucleus as a threadlike matrix called **chromatin.** When the cell begins to divide, the chromatin condenses into rod-shaped bodies called **chromosomes,** each of which, before dividing, is made up of two long DNA molecules and various histone molecules. The histones serve to organize the lengthy DNA, coiling it into bundles called nucleosomes. Also visible within the nucleus are one or more nucleoli, each consisting of RNA that is involved in the process of manufacturing the components of ribosomes. The components of ribosomes move to the cytoplasm to form a complete ribosome. The ribosome will eventually assemble amino acids into proteins. The nucleus also serves as the site for the separation of chromosomes during cell division.

■ The *endoplasmic reticulum,* or ER, consists of stacks of flattened sacs involved in the production of various materials. In cross-section, they appear as a series of mazelike channels, often closely associated with the nucleus. When ribosomes are present, the ER (called *rough ER*) attaches polysaccharide groups to polypeptides as they are assembled by the ribosomes. *Smooth ER,* without ribosomes, is responsible for various activities, including the synthesis of lipids and hormones, especially in cells that produce these substances for export from the cell. In liver cells, smooth ER is involved in the breakdown of toxins, drugs, and toxic byproducts from cellular reactions.

■ A *Golgi apparatus* (*Golgi complex* or *Golgi body*) is a group of flattened sacs arranged like a stack of bowls. They function to modify and package proteins and lipids into *vesicles,* small, spherically shaped sacs that bud from the ends of a Golgi apparatus. Vesicles often migrate to and merge with the plasma membrane, releasing their contents outside of the cell.

Figure 2-2 The general organization of a typical cell.

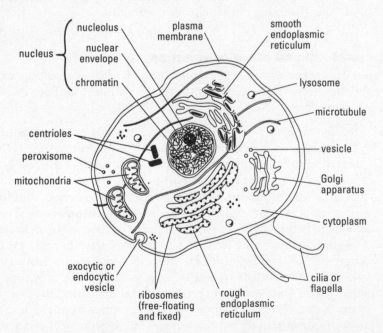

■ *Lysosomes* are vesicles from a Golgi apparatus that contain digestive enzymes. They break down food, cellular debris, and foreign invaders such as bacteria.

■ *Mitochondria* carry out aerobic respiration, a process in which energy (in the form of ATP) is obtained from carbohydrates. The mitochondria can also produce energy from noncarbohydrate sources such as fats.

■ *Ribosomes* carry out the process of producing protein.

■ *Vaults* are one of the newest organelles discovered. It appears they function to transport messenger RNA through the cytosol to the ribosomes. They seem to also be involved in developing drug resistance.

■ *Microtubules, intermediate filaments,* and *microfilaments* are three protein fibers of decreasing diameter, respectively. All are involved in establishing the shape or movements of the *cytoskeleton,* the internal structure of the cell.

Microtubules are made of the protein tubulin and provide support and mobility for cellular activities. They are found in the spindle apparatus (which guides the movement of chromosomes during cell division) and in flagella and cilia (described later in this list), which project from the plasma membrane to provide motility to the cell.

Intermediate filaments help support the shape of the cell.

Microfilaments are made of the protein actin and are involved in cell motility. They are found in almost every cell, but are predominant in muscle cells and in cells that move by changing shape, such as phagocytes (white blood cells that scour the body for bacteria and other foreign invaders).

■ *Flagella* and *cilia* protrude from the cell membrane and make wavelike movements. Flagella and cilia are classified by their lengths and by their number per cell: Flagella are long and few; cilia are short and many. A single flagellum propels sperm, while the numerous cilia that line the respiratory tract sweep away debris. Structurally, both flagella and cilia consist of microtubules arranged in a "9 + 2" array—that is, nine pairs (doublets) of microtubules arranged in a circle surrounding a pair of microtubules (Figure 2-3).

Figure 2-3 The structural arrangement of various cell specializations.

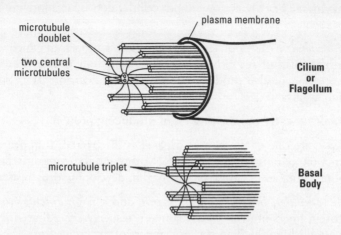

■ **Centrioles** and **basal bodies** act as microtubule organizing centers (MTOCs). A pair of centrioles (enclosed in a centrosome) located outside the nuclear envelope gives rise to the microtubules that make up the spindle apparatus used during cell division. Basal bodies are at the base of each flagellum and cilium and appear to organize their development. Both centrioles and basal bodies are made up of nine triplets arranged in a circle (Figure 2-3).

■ *Peroxisomes* are organelles common in liver and kidney cells that break down potentially harmful substances. Some chemical reactions in the body produce a byproduct called hydrogen peroxide. Peroxisomes can convert hydrogen peroxide (a toxin made of H_2O_2) to water and oxygen.

Cell Junctions

The plasma membranes of adjacent cells are usually separated by extracellular fluids that allow transport of nutrients and wastes to and from the bloodstream. In certain tissues, however, the membranes of adjacent cells may join and form a junction. As shown in Figure 2-4, three kinds of cell junctions are recognized:

Figure 2-4 The three types of cell junctions.

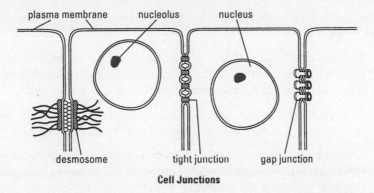

Cell Junctions

■ **Desmosomes** are protein attachments between adjacent cells. Inside the plasma membrane, a desmosome bears a disk-shaped structure from which protein fibers extend into the cytoplasm. Desmosomes act like spot welds to hold together tissues that undergo considerable stress (such as skin or heart muscle).

■ **Tight junctions** are tightly stitched seams between cells. The junction completely encircles each cell, preventing the movement of material between the cell. Tight junctions are characteristic of cells lining the digestive tract, where materials are required to pass through cells (rather than intercellular spaces) to penetrate the bloodstream.

■ **Gap junctions** are narrow tunnels between cells that consist of proteins called connexons. The proteins allow only the passage of ions and small molecules. In this manner, gap junctions allow communication between cells through the exchange of materials or the transmission of electrical impulses.

Movement of Substances

There are a few concepts that need to be understood relating to the movement of substances.

- The movement of substances may occur across a semi-permeable membrane (such as the plasma membrane). A semi-permeable membrane allows some substances to pass through, but not others.

- The substances, whose movements are being described, may be water (the solvent) or the substance dissolved in the water (the solute).

- Movement of substances may occur from higher to lower concentrations (down the concentration gradient) or from the opposite direction (up or against the gradient).

- Solute concentrations vary. A solution may be *hypertonic* (a higher concentration of solutes), *hypotonic* (a lower concentration of solutes), or *isotonic* (an equal concentration of solutes) compared to another region.

- The movement of substances may be passive or active. If movement is with the concentration or gradient, it is passive. If movement is against the gradient, it is active and requires energy.

Passive transport process

Passive transport describes the movement of substances down a concentration gradient and does not require energy consumption.

- *Diffusion* is the net movement of substances from an area of higher concentration to an area of lower concentration. This movement occurs as a result of the random and constant motion characteristic of all molecules, atoms, or ions (due to kinetic energy) and is independent from the motion of other molecules. Since at any one time some molecules may be moving against the concentration gradient and some molecules may be moving down the concentration gradient (remember, the motion is random), the word "net" is used to indicate the overall, eventual end result of the movement. If a concentration gradient exists, the molecules (which are constantly moving) will eventually become evenly distributed (a state of equilibrium).

■ **Osmosis** is the diffusion of water molecules across a semi-permeable membrane. When water moves into a cell by osmosis, hydrostatic pressure (osmotic pressure) may build up inside the cell.

■ **Dialysis** is the diffusion of solutes across a semi-permeable membrane.

■ **Facilitated diffusion** is the diffusion of solutes through channel proteins in the plasma membrane. Note that water can pass freely through the plasma membrane without the aid of specialized proteins, although special proteins called **aquaporins** can aid or speed-up water transport.

Active transport processes

Active transport is the movement of solutes against a gradient and requires the expenditure of energy (usually ATP). Active transport is achieved through one of the following two mechanisms:

■ Transport proteins in the plasma membrane transfer solutes such as small ions (Na^+, K^+, Cl^-, H^+), amino acids, and monosaccharides.

■ Vesicles or other bodies in the cytoplasm move macromolecules or large particles across the plasma membrane. Types of vesicular transport include the following:

■ *Exocytosis,* which describes the process of vesicles fusing with the plasma membrane and releasing their contents to the outside of the cell. This process is common when a cell produces substances for export.

■ *Endocytosis,* which describes the capture of a substance outside the cell when the plasma membrane merges to engulf it. The substance subsequently enters the cytoplasm enclosed in a vesicle. There are three kinds of endocytosis:

■ *Phagocytosis* ("cellular eating") occurs when undissolved material enters the cell. The plasma membrane engulfs the solid material, forming a phagocytic vesicle.

- *Pinocytosis* ("cellular drinking") occurs when the plasma membrane folds inward to form a channel allowing dissolved substances to enter the cell. When the channel is closed, the liquid is enclosed within a pinocytic vesicle.

- *Receptor-mediated endocytosis* occurs when specific molecules in the fluid surrounding the cell bind to specialized receptors in the plasma membrane. As in pinocytosis, the plasma membrane folds inward and the formation of a vesicle follows. Certain hormones are able to target specific cells by receptor-mediated endocytosis.

Cell Division

Cell division consists of two phases—*nuclear division* followed by *cytokinesis*. Nuclear division divides the genetic material in the nucleus, while cytokinesis divides the cytoplasm. There are two kinds of nuclear division—mitosis and meiosis. Mitosis divides the nucleus so that both daughter cells are genetically identical. In contrast, meiosis is a reduction division, producing daughter cells that contain half the genetic information of the parent cell.

The first step in either mitosis or meiosis begins with the condensation of the genetic material, chromatin, into tightly coiled bodies, the **chromosomes.** Each chromosome is made of two identical halves called sister **chromatids,** which are joined at the **centromere.** Each chromatid consists of a single, tightly coiled molecule of DNA. Somatic cells (all body cells except eggs and sperm) are diploid cells because each cell contains two copies of every chromosome. A pair of such chromosomes is called a homologous pair. In a **homologous pair of chromosomes,** one homologue originates from the maternal parent, the other from the paternal parent. In humans there are 46 chromosomes (23 homologous pairs). In males there are only 22 homologous pairs (autosomes) and one nonhomologous pair—the sex chromosomes of X and Y.

When a cell is not dividing, the chromatin is enclosed within a clearly defined nuclear envelope, one or more nucleoli are visible within the nucleus, and two centrosomes (each containing two centrioles) lie adjacent to one another outside the nuclear envelope. These features are characteristic of *interphase,* the nondividing but metabolically active period of the cell cycle (Figure 2-5). When cell division begins, these features change, as described in the following sections.

Figure 2-5 Stages of the cell cycle.

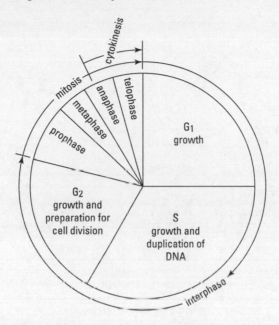

Mitosis

There are four phases in **mitosis** (adjective, mitotic): prophase, metaphase, anaphase, and telophase (Figure 2-6):

■ During *prophase,* the nucleoli disappear, the chromatin condenses into chromosomes, the nuclear envelope breaks down, and the mitotic spindle is assembled. The development of the mitotic spindle begins as the centrosomes move apart to opposite ends (poles) of the nucleus. As they move apart, microtubules develop from each centrosome, increasing in length by the addition of tubulin units. Microtubules from each centrosome connect to specialized regions in the centromere called **kinetochores.** Microtubules tug on the kinetochores, moving the chromosomes back and forth toward one pole, then the other. Within the spindle, there are also microtubules that overlap at the center of the spindle and do not attach to the chromosomes.

Figure 2-6 Cell reproduction and the four stages of mitosis.

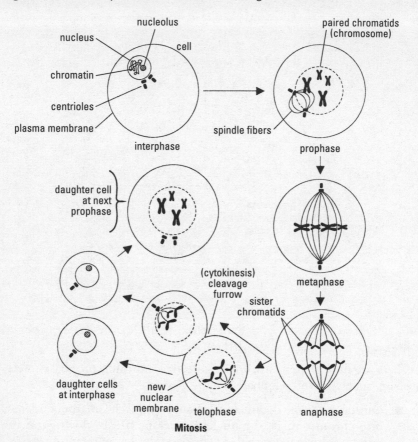

Mitosis

■ *Metaphase* begins when the chromosomes are distributed across the metaphase plate, a plane lying between the two poles of the spindle. Metaphase ends when the microtubules, still attached to the kinetochores, pull each chromosome apart into two chromatids. Each chromatid is complete with a centromere and kinetochores. Once separated from its sister chromatid, each chromatid is called a chromosome. (To count the number of chromosomes at any one time, count the number of centromeres.)

- *Anaphase* begins after the chromosomes are separated into individual chromatids. During anaphase, the microtubules connected to the chromatids (now chromosomes) shorten, effectively pulling the chromosomes to opposite poles. Overlapping microtubules, originating from opposite centrosomes but not attached to chromosomes, interact to push the poles farther apart. At the end of anaphase, each pole has a complete set of chromosomes, the same number of chromosomes as the original cell. (Since it consists of only one chromatid, each chromosome contains only a single copy of the DNA molecule.)

- *Telophase* concludes the nuclear division. During this phase, a nuclear envelope develops around each pole, forming two nuclei. The chromosomes within each of these nuclei disperse into chromatin, and the nuclei reappear. Simultaneously, cytokinesis occurs, dividing the cytoplasm into two cells. Microfilaments form a ring inside the plasma membrane between the two newly forming nuclei. As the microfilaments shorten, they act like purse strings to pull the plasma membrane into the center, dividing the cell into two daughter cells. The groove that forms as the purse strings are tightened is called a *cleavage furrow*.

Once mitosis is completed and interphase begins, the cell begins a period of growth. Growth begins during the first phase, called G_1 (gap), and continues through the S (synthesis) and G_2 phases. Also during the S phase the second DNA molecule for each chromosome is synthesized. As a result of this DNA replication, each chromosome gains a second chromatid. During the G_2 period of growth, materials for the next mitotic division are prepared. The time span from one cell division through G_1, S, and G_2 is called a *cell cycle* (Figure 2-5).

A cell that begins mitosis in the diploid state—that is, with two copies of every chromosome—will end mitosis with two copies of every chromosome. However, each of these chromosomes will consist of only one chromatid, or one DNA molecule. During interphase, the second DNA molecule is replicated from the first, so that when the next mitotic division begins, each chromosome will again consist of two chromatids.

Meiosis

Meiosis (adjective, meiotic) is very similar to mitosis. The major distinction is that meiosis consists of two groups of divisions, meiosis I and meiosis II (Figure 2-7). In meiosis I, homologous chromosomes pair at the metaphase plate and then migrate to opposite poles. In meiosis II, chromosomes spread across the metaphase plate, and sister chromatids separate and migrate to opposite poles. Thus, meiosis II is analogous to mitosis. A summary of each meiotic stage follows:

- Prophase I begins like prophase of mitosis. The nucleolus disappears, chromatin condenses into chromosomes, the nuclear envelope breaks down, and the spindle apparatus develops. Once the chromosomes are condensed, however, their behavior differs from mitosis. During prophase I, homologous chromosomes pair, a process called *synapsis*. These pairs of homologous chromosomes are called *tetrads* (a group of four chromatids) or bivalents. During synapsis, corresponding regions form close associations called **chiasmata** (singular, chiasma) along nonsister chromatids. Chiasmata are sites where genetic material is exchanged between nonsister homologous chromatids, a process called *crossing over*. The result contributes to a mixing of genetic material from both parents, a process called genetic recombination.

- At metaphase I, homologous pairs of chromosomes are spread across the metaphase plate. Microtubules extending from one pole are attached to kinetochores of one member of each homologous pair. Microtubules from the other pole are connected to the second member of each homologous pair.

- Anaphase I begins when homologues within tetrads uncouple as they are pulled to opposite poles.

- In telophase I, the chromosomes have reached their respective poles, and a nuclear membrane develops around them. Note that each pole will form a new nucleus that will have half the number of chromosomes, but each chromosome will contain two chromatids. Since daughter nuclei will have half the number of chromosomes, cells that they eventually form will be haploid.

- Cytokinesis occurs, forming two daughter cells. A brief interphase may follow, but no replication of chromosomes occurs. Instead, part II of meiosis begins in both daughter nuclei.

■ In prophase II, the nuclear envelope disappears and the spindle develops. There are no chiasmata and no crossing over of genetic material as in prophase I.

Figure 2-7 The stages of meiosis.

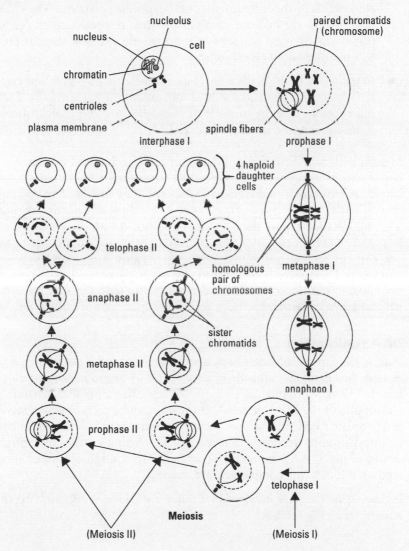

■ In metaphase II, the chromosomes align singly on the metaphase plate (not in tetrads as in metaphase I). Single alignment of chromosomes is exactly what happens in mitosis—except now there is only half the number of chromosomes.

■ Anaphase II begins as each chromosome is pulled apart into two chromatids by the microtubules of the spindle apparatus. The chromatids (now chromosomes) migrate to their respective poles. Again, this is exactly what happens in mitosis—except now there is only half the number of chromosomes.

■ In telophase II, the nuclear envelope reappears at each pole and cytokinesis occurs. The end result of meiosis is four haploid cells. Each cell contains half the number of chromosomes and each chromosome consists of only one chromatid.

Meiosis ends with four haploid daughter cells, each with half the number of chromosomes (one chromosome from each homologous pair). These are **gametes**—that is, eggs and sperm. The fusing of an egg and sperm, fertilization (*syngamy*), gives rise to a diploid cell, the **zygote.** The single-celled zygote then divides by mitosis to produce a multicellular embryo fetus, and after nine months, a newborn infant. Note that one copy of each chromosome pair in the zygote originates from one parent, and the second copy from the other parent. Thus, a pair of homologous chromosomes in the diploid zygote represents both maternal and paternal heritage.

DNA replication

During the S phase of interphase, a second chromatid is assembled. The second chromatid contains the exact same DNA found in the first chromatid. The copying process, called *DNA replication,* involves separating ("unzipping") the DNA molecule into two strands, each of which serves as a template to assemble a new, complementary strand. The result is two identical double-stranded molecules of DNA that consist of a single strand of old DNA (the template strand) and a single strand of new, replicated DNA (the complementary strand).

Following are the steps involved in duplicating DNA. While studying the steps, refer to Figure 2-8:

Figure 2-8 DNA Replication.

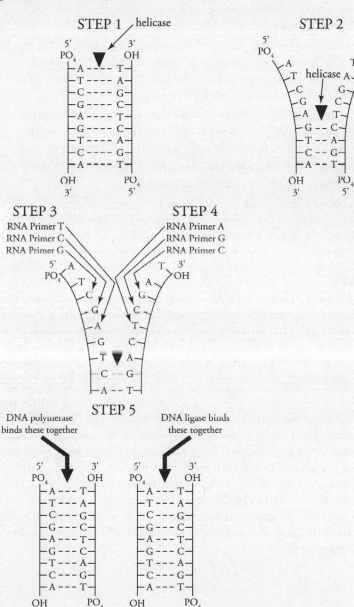

1. Each strand of DNA is labeled as 3' and 5'. The 3' area terminates with a hydroxyl group and the 5' area terminates with a phosphate group.

2. The enzyme helicase "unzips" (unwinds) the DNA helix, producing a Y-shaped replication fork. Note: The DNA shown in Figure 2-8 is not depicted in a helical shape; it is drawn in a parallel form for ease of understanding.

3. RNA primers "bring" in respective base pairings to each of the original strands. DNA polymerase is an enzyme that binds the base pairings together, but it can only work in the direction of 5' to 3'.

4. The other original strand also has to be "put together" 5' to 3' so it will be put together in a backward fashion.

5. In order to bind those base pairings to the original strand, a different enzyme called *DNA ligase* is necessary. This is called the "lagging strand" since it basically takes longer to put together.

Mutations

The replication process of DNA is extremely accurate; however, errors can occur when nucleotide bases between DNA strands are occasionally paired incorrectly. In addition, errors in DNA molecules may arise as a result of exposure to radiation (such as ultraviolet or X-ray) or various reactive chemicals. When errors occur, repair mechanisms are available to make corrections.

If a DNA error is not repaired, it becomes a mutation. A **mutation** is any sequence of nucleotides in a DNA molecule that does not exactly match the original DNA molecule from which it was copied. Mutations include an incorrect nucleotide (substitution), a missing nucleotide (deletion), or an additional nucleotide not present in the original DNA molecule (insertion). When an insertion mutation occurs, it causes all subsequent nucleotides to be displaced one position, producing a frameshift mutation. Radiation or chemicals that cause mutations are called **mutagens.** Mutagens that activate uncontrolled cell growth (cancer) are called **carcinogens.**

Protein synthesis

The DNA in chromosomes contains genetic instructions that regulate development, growth, and the metabolic activities of cells. The DNA instructions determine whether a cell will be that of a pea plant, a human, or some other organism, as well as establish specific characteristics of the cell in that organism. For example, the DNA in a cell may establish that it is a human cell. If, during development, it becomes a cell in the iris of any eye, the DNA will direct other information appropriate for its location in the organism, such as the production of brown, blue, or other pigmentation. DNA controls the cell in this manner because it contains codes for polypeptides. Many polypeptides are enzymes that regulate chemical reactions and influence the resulting characteristics of the cell. Thus, from the molecular viewpoint, traits are the end products of metabolic processes regulated by enzymes. A *gene* is defined as the DNA segment that codes for a particular enzyme or other polypeptide (one-gene-one-polypeptide hypothesis).

The process that describes how enzymes and other proteins are made from DNA is called *protein synthesis*. There are three steps in protein synthesis: transcription, RNA processing, and translation. In transcription, DNA molecules are used as a template to create RNA. After transcription, RNA processing modifies the RNA molecule with deletions and additions. In translation, the processed RNA molecules are used to assemble amino acids into a polypeptide.

There are three kinds of RNA molecules produced during transcription:

- **Messenger RNA (mRNA)** is a single strand of RNA that provides the template used for sequencing amino acids into a polypeptide. A triplet group of three adjacent nucleotides on the mRNA, called a **codon,** codes for one specific amino acid. There are 64 possible ways that four nucleotides can be arranged in triplet combinations ($4 \times 4 \times 4 = 64$ possible codons). The genetic code is a table of information that provides "decoding" for each codon—that is, it identifies the amino acid specified by each of the possible 64 codon combinations. For example, the codon composed of the three nucleotides cytosine-guanine-adenine (CGA) codes for the amino acid arginine. Refer to Figure 2-9.

Figure 2-9 Combinations of Amino Acid Production.

Universal Codon Chart

First Letter	Second Letter				Third Letter
	U	C	A	G	
U	phenylalanine	serine	tyrosine	cysteine	U
	phenylalanine	serine	tyrosine	cysteine	C
	leucine	serine	stop	stop	A
	leucine	serine	stop	tryptophan	G
C	leucine	proline	histidine	arginine	U
	leucine	proline	histidine	arginine	C
	leucine	proline	glutamine	arginine	A
	leucine	proline	glutamine	arginine	G
A	isoleucine	threonine	asparagine	serine	U
	isoleucine	threonine	asparagine	serine	C
	isoleucine	threonine	lysine	arginine	A
	(start) methionine	threonine	lysine	arginine	G
G	valine	alanine	aspartate	glycine	U
	valine	alanine	aspartate	glycine	C
	valine	alanine	glutamate	glycine	A
	valine	alanine	glutamate	glycine	G

How to use the chart:

1. The code for the production of the amino acid leucine is CUA, CUG, CUC, CUU, UUA, or UUG.

2. The code for the production of the amino acid lysine is AAA or AAG.

3. The code for the amino acid cysteine is UGU or UGC.

■ Transfer RNA (tRNA) is a short RNA molecule (consisting of about 80 nucleotides) that is used for transporting amino acids to their proper places on the mRNA template. Interactions among various parts of the tRNA molecule result in base-pairings between nucleotides, folding the tRNA in such a way that it forms a three-dimensional molecule. (In two dimensions, tRNA resembles the three parts of a clover leaf.) One end of the tRNA attaches to an amino acid. Another portion of the tRNA, specified by a triplet combination of nucleotides, is the anticodon. During translation, the anticodon of the tRNA base pairs with the codon of the mRNA.

■ Ribosomal RNA (rRNA) molecules are the building blocks of ribosomes. The nucleolus is an assemblage of DNA actively being transcribed into rRNA. Within the nucleolus, various proteins imported from the cytosol are assembled with rRNA to form large and small ribosome subunits. Together, the two subunits form a ribosome, which coordinates the activities of the mRNA and tRNA during translation. Ribosomes have three binding sites—one for the mRNA, one for the tRNA that carries a growing polypeptide chain, and one for a second tRNA that delivers the next amino acid that will be inserted into the growing polypeptide chain.

Here are the details of transcription, RNA processing, and protein synthesis (also see Figures 2-10 and 2-11):

■ During transcription, the RNA polymerase attaches to promoter regions on the DNA and beings to unzip the DNA into two strands. (See Step 1 in Figure 2-10.)

■ As the RNA polymerase unzips the DNA, it assembles new nucleotides using one strand of the DNA as a template. In contrast to the process of DNA replication, the new nucleotides are RNA nucleotides, and only one DNA strand is transcribed. (See Step 2 in Figure 2-10.)

■ Transcription continues until the RNA polymerase reaches a special sequence of nucleotides that serves as a termination point. The RNA polymerase and the newly created RNA molecule are released. This newly created RNA molecule may be mRNA, tRNA, or rRNA, depending on which DNA segment is transcribed. (See Step 3 in Figure 2-10.)

Figure 2-10 Transcription and RNA processing.

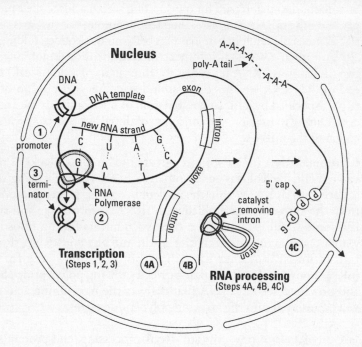

■ During RNA processing, newly created mRNA molecules undergo two kinds of alterations. In the first modification, noncoding intervening sequences called **introns** are removed, leaving only **exons,** sequences that express a code for a polypeptide. A second modification adds two special sequences—a 5-inch cap to one end of the mRNA and a poly-A tail to the other end. (See steps 4A, 4B, and 4C in Figure 2-10.)

■ The mRNA, tRNA, and ribosomal subunits are transported across the nuclear envelope and into the cytoplasm. In the cytoplasm, amino acids attach to one end of the tRNAs. (See steps 5A, 5B, and 5C in Figure 2-11.)

Figure 2-11 The steps involved in protein synthesis.

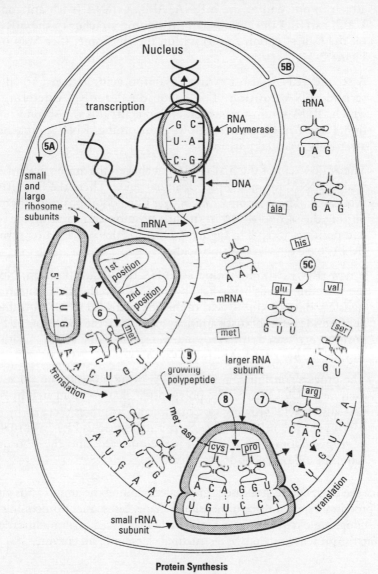

Protein Synthesis

- Translation begins when the small and large ribosomal subunits attach to one end of the mRNA. Also, a tRNA (with anticodon UAC) carrying the amino acid methonine attaches to the mRNA (at the "start" codon AUG) within the ribosome. (See Step 6 in Figure 2-11.)

- A second tRNA, also bearing an amino acid, arrives and fills a second tRNA position. The codon on the mRNA determines which tRNA (and thus, which amino acid) fills the second position. (Step 7 in Figure 2-11 shows an incoming tRNA approaching a yet-to-be-vacated position.)

- The amino acid of the first tRNA attaches to the amino acid of the second tRNA, forming a pair of amino acids. Then, the first tRNA is released. The ribosome moves over one codon position, thereby putting the second tRNA in the first position and vacating the second position. (Step 8 in Figure 2-11 shows this process after several tRNAs have delivered amino acids.)

- A new tRNA (with its amino acid) fills the vacant position. Now, the two amino acids being held by the tRNA in the first position are transferred to the amino acid of the newly arrived tRNA, forming a polypeptide chain of three amino acids. Again, the tRNA in the first position is released, the ribosome moves over one codon position, and the second tRNA position is vacant.

- The process continues, as new tRNAs bring more amino acids. As each new tRNA arrives, the polypeptide chain is elongated by one new amino acid, growing in sequence and length as dictated by the codons on the mRNA. (See Step 9 in Figure 2-11.) Eventually, a "stop" codon, such as UAG, is encountered, and the ribosome subunits and polypeptide are released.

Once the polypeptide is released, interactions among the amino acids gives the protein its special three-dimensional shape. Subsequent processing by the endoplasmic reticulum or a Golgi body may make final modifications before the protein functions as a structural element or an enzyme.

Chapter Check-Out

Q&A

1. _____ provide support and motility to a cell while projecting from the plasma membrane.

2. True or False: Gap junctions allow the transmission of electrical impulses and exchange of material between neighboring cells.

3. In mitosis, which of the following steps is not a characteristic of telophase?

    a. Chromosomes disperse into chromatin.
    b. Microtubules shorten and pull chromosomes to opposite poles.
    c. A cleavage furrow is visible.
    d. The cytoplasm begins dividing into two cells.

4. True or False: DNA polymerase initiates replication by itself via the utilization of DNA nucleotides.

5. During _____, DNA is unzipped into two separate strands by RNA polymerase.

Answers: 1. Microtubules, **2.** T, **3.** b, **4.** F, **5.** transcription

Chapter 3

TISSUES

Chapter Check-In

❑ Discovering the four basic types of tissue

❑ Listing the general characteristics and functions of the different types of epithelium

❑ Describing the various kinds of connective tissue and their components

❑ Understanding the different kinds of nervous tissue and muscle

The human body is composed of approximately 200 distinctly different types of cells. These cells are organized into four basic tissues that, in turn, are assembled to form organs. When you examine tissue at a microscopic level, having the ability to detect the presence and location of the four basic tissues enables you to identify the organ that you are looking at. A basic knowledge of the general characteristics and cellular composition of these tissues is essential in *histology*, which is the study of tissues at the microscopic level.

Tissues are groups of similar cells performing a common function. There are four categories of tissues:

- Epithelial tissue
- Connective tissue
- Nervous tissue
- Muscle tissue

Epithelial Tissue

Epithelial tissue, or epithelium, has the following general characteristics:

- Epithelium consists of closely packed, flattened cells that make up the inside or outside lining of body areas. There is little intercellular material.

- The tissue is **avascular,** meaning without blood vessels. Nutrient and waste exchange occurs through neighboring connective tissues by diffusion.

- The upper surface of epithelium is free, or exposed to the outside of the body or to an internal body cavity. The basal surface rests on connective tissue. A thin, extracellular layer called the *basement membrane* forms between the epithelial and connective tissue.

There are two kinds of epithelial tissues:

- Covering and lining epithelium covers the outside surfaces of the body and lines internal organs.

- Glandular epithelium secretes hormones or other products.

Epithelium that covers or lines

Epithelial tissues that cover or line surfaces are classified by cell shape and by the number of cell layers. The following terms are used to describe these features.

Cell shape:

- *Squamous cells* are flat. The nucleus, located near the upper surface, gives these cells the appearance of a fried egg.

- *Cuboidal cells* are cube- or hexagon-shaped with a central, round nucleus. These cells produce secretions (sweat, for example) or absorb substances such as digested food.

- *Columnar cells* are tall with an oval nucleus near the basement membrane. These thick cells serve to protect underlying tissues or may function to absorb substances. Some have microvilli, minute surface extensions, to increase surface area for absorbing substances, while others may have cilia that help move substances over their surface (such as mucus through the respiratory tract).

- *Transitional cells* range from flat to tall cells that can extend or compress in response to body movement.

Number of cell layers:

- *Simple epithelium* describes a single layer of cells.

- *Stratified epithelium* describes epithelium consisting of multiple layers.

- *Pseudostratified epithelium* describes a single layer of cells of different sizes, giving the appearance of being multilayered.

Names of epithelial tissues include a description of both their shape and their number of cell layers. The presence of cilia may also be identified in their names. For example, simple squamous describes epithelium consisting of a single layer of flat cells. Pseudostratified columnar ciliated epithelium describes a single layer of tall, ciliated cells of more than one size. Stratified epithelium is named after the shape of the outermost cell layer. Thus, stratified squamous epithelium has outermost layers of squamous cells, even though some inner layers consist of cuboidal or columnar cells. These and other epithelial tissues are illustrated in Figure 3-1.

Figure 3-1   Types of epithelial tissues.

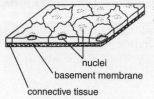

simple squamous epithelium

Cells: single (scalelike) layer
Nuclei: flattened, centrally located
Functions: diffusion, lubrication

nuclei
basement membrane
connective tissue

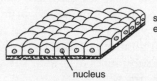

simple cuboidal epithelium

Cells: single (squarelike) layer
Nuclei: centrally located
   and spherical
Functions: absorption,
   secretion, protection

nucleus

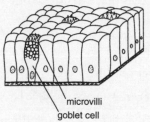

simple columnar epithelium

Cells: tall, single-layered
Nuclei: basally located and
   enlongated
Functions: absorption,
   secretion, protection
(May bear cilia and may contain
   goblet cells with microvilli)

microvilli
goblet cell

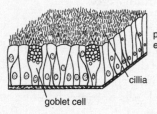

pseudostratified epithelium

Cells: differ in height, not all cells
   reach the apical surface
Nuclei: at various positions
Functions: absorption, secretion,
   transportation

cilia
goblet cell

Epithelial Tissues

stratified squamous
epithelium

Cells: squamous cells apically,
but basal layers vary from
cuboidal to columnar
Nuclei: centrally located
Functions: protection

cuboidal cells
squamous cells

Cells: two layers
Nuclei: centrally located
and spherical
Functions: absorption,
secretion

stratified cuboidal
epithelium

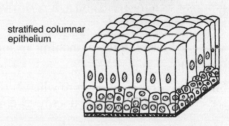

connective tissue
basement membrane

Cells: single layer of
columnar cells on
several layers of
cuboidal (or many
sided) cells
Nuclei: basal and oval
Functions: protection,
secretion

stratified columnar
epithelium

Cells: vary depending on
stretch, apical cells often
large, round, and bi-
nucleated
Nuclei: centrally located
Functions: distention
(occurs only in
bladder, ureter, and
urethra)

transitional
epithelium

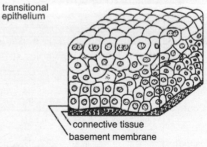

connective tissue
basement membrane

Epithelial Tissues

Glandular epithelium

Glandular epithelium forms two kinds of glands:

- *Endocrine glands* secrete **hormones** directly into the bloodstream. For example, the thyroid gland secretes the hormone thyroxin into the bloodstream, where it is distributed throughout the body, stimulating an increase in the metabolic rate of body cells.

- *Exocrine glands* secrete their substances into tubes, or ducts, which carry the secretions to the epithelial surface. Examples of secretions include sweat, saliva, milk, stomach acid, and digestive enzymes.

Exocrine glands are classified according to their structure (see Figure 3-2):

- Unicellular or multicellular describes a single-celled gland or a gland made of many cells, respectively. A multicellular gland consists of a group of secretory cells and a duct through which the secretions pass as they exit the gland.

- Branched refers to the branching arrangement of secretory cells in the gland.

- Simple or compound refers to whether the duct of the gland (not the secretory portion) does or does not branch, respectively.

- Tubular describes a gland whose secretory cells form a tube, while alveolar (or acinar) describes secretory cells that form a bulblike sac.

Figure 3-2 Exocrine glands can be classified as simple or compound with either a tubular or alveolar structure.

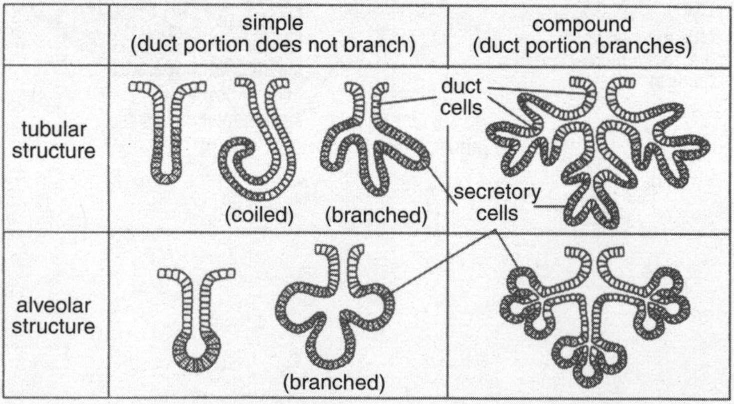

Exocrine glands are also classified according to their function (see Figure 3-3):

- In merocrine glands, secretions pass through the cell membranes of the secretory cells (exocytosis). For example, goblet cells of the trachea release mucus via exocytosis.

- In apocrine glands, a portion of the cell containing secretions is released as it separates from the rest of the cell. For example, the apical portion of lactiferous glands release milk in this manner.

- In holocrine glands, entire secretory cells disintegrate and are released along with their contents. For example, sebaceous glands release sebum to lubricate the skin in this manner.

Figure 3-3 Exocrine glands can be classified according to their function.

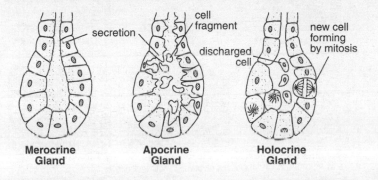

Connective Tissue

A summary of the various kinds of connective tissues is given in Figure 3-4 and Table 3-1.

Figure 3-4 General characteristics of connective tissues.

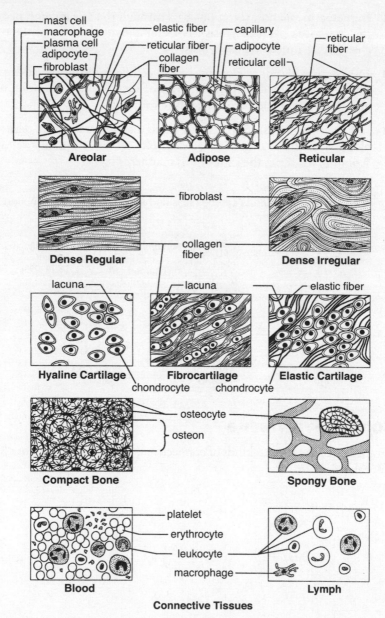

Connective Tissues

Table 3-1 Kinds of Connective Tissue

Tissue Type	Cells Present	Fibers Present	Matrix Characteristics
Loose connective tissue			
areolar	fibroblasts, macrophages, adipocytes, mast cells, plasma cells	collagen, elastic, reticular	loosely arranged fibers in gelatinous ground substance
adipose	adipocytes	reticular, collagen	closely packed cells with a small amount of gelatinous ground substance; stores fat
reticular	reticular cells	reticular	loosely arranged fibers in gelatinous ground substance
Dense connective tissue			
dense regular	fibroblasts	collagen	parallel-arranged bundles of fibers with few cells and little ground substance; great tensile strength
dense irregular	fibroblasts	collagen (some elastic)	irregularly arranged bundles of fibers with few cells and little ground substance; high tensile strength
Cartilage			
hyaline	chondrocytes	collagen (some elastic)	limited ground substance; dense, gel matrix
fibrocartilage	chondrocytes	collagen (some elastic)	limited ground substance intermediate between hyaline cartilage and dense connective tissue
elastic	chondrocytes	elastic	limited ground substance; flexible but firm gel matrix

(continued)

Table 3-1 *(continued)*

Tissue Type	Cells Present	Fibers Present	Matrix Characteristics
Bone (osseous tissue)			
compact (dense)	osteoblasts, osteocytes	collagen	rigid, calcified ground substance with osteons
spongy (cancellous)	osteoblasts, osteocytes	collagen	rigid, calcified ground substance (no osteons)
Blood and lymph (vascular tissue)			
blood	erythrocytes, leukocytes, platelets	"fibers" are soluble proteins that form during clotting	"matrix" is liquid blood plasma
lymph	leukocytes	"fibers" are soluble proteins	"matrix" is liquid lymph

The following information identifies a few select features of connective tissue.

- *Nerve supply.* Most connective tissues have a nerve supply (as does epithelial tissue).

- *Blood supply.* There is a wide range of vascularity among connective tissues, although most are well vascularized (unlike epithelial tissues, which are all avascular).

- *Structure.* Connective tissue consists of scattered cells immersed in an intercellular material called the matrix. The matrix consists of fibers and ground substance. The kinds and amounts of fiber and ground substance determine the character of the matrix, which in turn defines the kind of connective tissue.

- *Cell types.* Fundamental cell types, characteristic of each kind of connective tissue, are responsible for producing the matrix. Immature forms of these cells (whose names end in *blast*) secrete the fibers and ground substance of the matrix. Cells that have matured, or differentiated (whose names often end in *cyte*), function mostly to maintain the matrix:

  - Fibroblasts are common in both loose and dense connective tissues.

  - Adipocytes, cells that contain molecules of fat, occur in loose connective tissue, as does adipose tissue.

- Reticular cells resemble fibroblasts, but have long, cellular processes (extensions). They occur in loose connective tissue.

- Chondroblasts and chondrocytes occur in cartilage.

- Osteoblasts and osteocytes occur in bone.

- Hemocytoblasts occur in the bone marrow and produce erythrocytes (red blood cells), leukocytes (white blood cells), and platelets (formerly called thrombocytes).

In addition to the fundamental cell types, various leukocytes migrate from the bone marrow to connective tissues and provide various body defense activities:

- Macrophages engulf foreign and dead cells.

- Mast cells secrete histamine, which stimulates immune responses.

- Plasma cells produce antibodies.

- *Fibers.* Matrix fibers are proteins that provide support for the connective tissue. There are three types:

  - Collagen fibers, made of the protein collagen, are both tough and flexible.

  - Elastic fibers, made of the protein elastin, are strong and stretchable.

  - Reticular fibers, made of thin collagen fibers with a glycoprotein coating, branch frequently to form a netlike (reticulate) pattern.

- *Ground substance.* Ground substance may be fluid, gel, or solid, and, except for blood, is secreted by the cells of the connective tissue:

  - Cell adhesion proteins hold the connective tissue together.

  - Proteoglycans provide the firmness of the ground substance. Hyaluronic sulfate and chondroitin sulfate are two examples.

- *Classification.* There are five general categories of mature connective tissue:

  - Loose connective tissue has abundant cells among few or loosely arranged fibers and a sparse to abundant gelatinous ground substance.

  - Dense connective tissue has few cells among a dense network of fibers with little ground substance.

- Cartilage has cells distributed among fibers in a firm gellike ground substance. Cartilage is tough but flexible, avascular, and without nerves.

- Bone has cells distributed among abundant fibers in a solid ground substance containing minerals, mostly calcium phosphate. Bone is organized in units, called osteons (formerly known as the Haversian system). Each **osteon** consists of a central canal, which contains blood vessels and nerves, surrounded by concentric rings (lamellae) of hard matrix and collagen fibers. Branching off the central canal at right angles are perforating canals. These canals consist of blood vessels that branch off the central vessels (see Figure 5-1). Between the lamellae are cavities (lacunae) that contain bone cells (osteocytes). Canals (canaliculi) radiate from the central canal and allow nutrient and waste exchange with the osteocytes.

- Blood is composed of various blood cells and cell fragments (platelets) distributed in a fluid matrix called blood plasma.

- *Tissue origin.* All mature connective tissues originate from embryonic connective tissue. There are two kinds of embryonic connective tissues:

  - Mesenchyme is the origin of all mature connective tissues.

  - Mucous connective tissue is a temporary tissue formed during embryonic development.

An **epithelial membrane** is a combination of epithelial and connective tissues working together to perform a specific function. As such, it acts as an organ. There are four principle types of epithelial membranes:

- *Serous membranes* line interior organs and cavities. The serous membranes that line the heart, lungs, and abdominal cavities and organs are called the pericardium, pleura, and peritoneum, respectively.

- *Mucous membranes* line body cavities that open to the outside of the body. These include the nasal cavity and the digestive, respiratory, and urogenital tracts.

- *Synovial membranes* line the cavities at bone joints.

- The *cutaneous membrane* is the skin.

Nervous Tissue

Nervous tissue consists of two kinds of nerve cells:

- Neurons are the basic structural unit of the nervous system. Each cell consists of the following parts (see Figure 3-5):

    - The cell body contains the nucleus and other cellular organelles.

    - The dendrites are typically short, slender extensions of the cell body that receive stimuli.

    - The axon is typically a long, slender extension of the cell body that sends stimuli.

    - The axon branches are, typically, smaller extensions of the axon.

- Neuroglia, or glial cells, provide support functions for the neurons, such as insulation or anchoring neurons to blood vessels.

Figure 3-5 A neuron is a basic structural unit of the nervous system containing a cell body, dendrites, and an axon.

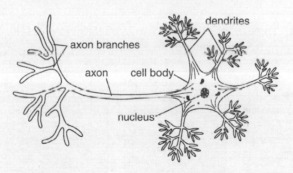

Muscle Tissue

There are three kinds of muscle tissues (see Figure 3-6):

- *Skeletal muscle* consists of long cylindrical cells that, under a microscope, appear striated with bands perpendicular to the length of the cell. The many nuclei in each cell (multinucleated cells) are located near the outside along the plasma membrane, which is called the sarcolemma. Skeletal muscle is attached to bones and causes movements of the body. Because it is under conscious control, it is also called *voluntary muscle.*

- *Cardiac muscle,* like skeletal muscle, is striated. However, cardiac muscle cells have a single, centrally located nucleus, and the muscle fibers branch often. Where two cardiac muscle cells meet, they form an intercalated disc containing gap junctions, which bridge the two cells. Cardiac cells are the only cells that pulsate in rhythm.

- *Smooth muscle* consists of cells with a single, centrally located nucleus. The cells are elongated with tapered ends and do not appear striated. Smooth muscle lines the walls of blood vessels and certain organs such as the digestive and urogenital tracts, where it serves to advance the movement of substances. Smooth muscle is called *involuntary muscle* because it is not under direct conscious control.

Figure 3-6 Three kinds of muscle tissue exist: skeletal muscle, cardiac muscle, and smooth muscle.

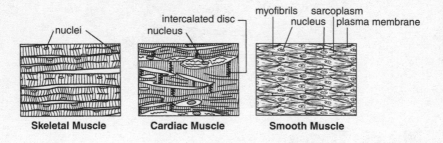

Skeletal Muscle   Cardiac Muscle   Smooth Muscle

Chapter Check-Out

Q&A

1. _____ are a type of epithelial specialization that greatly increases surface area to enhance absorption.

2. Which of the following statements is true concerning skeletal muscle?
   a. It has centrally located nuclei.
   b. It forms an intercalated disk between two cells.
   c. It lines the walls of the digestive tract.
   d. It contains multinucleated cells.

3. True or False: Blood is a type of connective tissue.

Answers: 1. Microvilli, **2.** d, **3.** T

Chapter 4

THE INTEGUMENTARY SYSTEM

Chapter Check-In

❏ Identifying the individual layers of each component of the skin

❏ Distinguishing the variety of accessory organs that reside within the skin

❏ Comprehending the physiological functions of the skin

The skin is far more than just the outer covering of human beings; it is an organ just like the heart, lung, or liver. Besides providing a layer of protection from pathogens, physical abrasions, and radiation from the sun, the skin serves many functions. It plays a vital role in homeostasis by maintaining a constant body temperature via the act of sweating or shivering and by making you aware of external stimuli through information perceived within the touch receptors located within the integumentary system. It only takes one visit to a burn unit to see the value of skin and the many complications that arise when this organ is compromised. In this chapter, you see the diverse actions of the integumentary system, as well as the composition of the skin that allows it to perform these many functions.

The Skin and Its Functions

The skin, or integument, is considered an organ because it consists of all four tissue types. The skin also consists of accessory organs, such as glands, hair, and nails, thus making up the integumentary system. A section of skin with various accessory organs is shown in Figure 4-1.

Figure 4-1   A section of skin with various accessory organs.

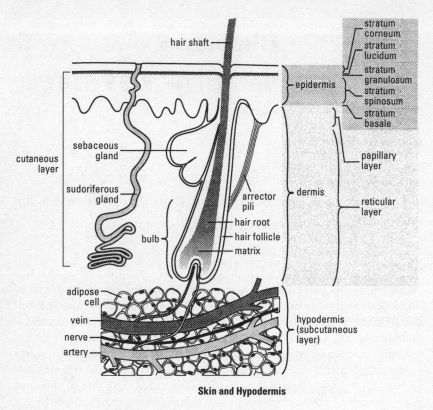

Skin and Hypodermis

The skin consists of two layers, the epidermis and the underlying dermis. Although technically not part of the skin, the hypodermis (subcutaneous layer, or superficial fascia) lies beneath the dermis.

The skin performs a variety of functions:

- Protection is provided against biological invasion, physical damage, and ultraviolet radiation.

- Sensation is provided by nerve endings for touch, pain, and heat.

- Thermoregulation is supported through the sweating and regulation of blood flow through the skin.

- Synthesis of vitamin D occurs in the skin.

- Blood within the skin can be shunted to other parts of the body when needed.

- Excretion of salts and small amounts of wastes (ammonia and urea) occurs with the production of sweat.

The Epidermis

The epidermis consists of keratinized stratified squamous epithelium. Four cell types are present:

- *Keratinocytes* produce keratin, a protein that hardens and water proofs the skin. Mature keratinocytes at the skin surface are dead and filled almost entirely with keratin.

- *Melanocytes* produce melanin, a pigment that protects cells from ultraviolet radiation. Melanin from the melanocytes is transferred to the keratinocytes.

- *Langerhans cells* are phagocytic macrophages that interact with white blood cells during an immune response.

- *Merkel cells* occur deep in the epidermis at the epidermal-dermal boundary. They form Merkel discs, which, in association with nerve endings, serve a sensory function.

There are several layers making up the epidermis. "Thick skin," found on the palms of the hands and soles of the feet, consists of five layers while "thin skin" consists of only four layers. Below is a list of all five layers:

1. The *stratum corneum* contains many layers of dead, anucleate keratinocytes completely filled with keratin. The outermost layers are constantly shed.

2. The *stratum lucidum* contains two to three layers of anucleate cells. This layer is found only in "thick skin" such as the palm of the hand and the sole of the foot.

3. The *stratum granulosum* contains two to four layers of cells held together by desmosomes. These cells contain keratohyaline granules, which contribute to the formation of keratin in the upper layers of the epidermis.

4. The *stratum spinosum* contains eight to ten layers of cells connected by desmosomes. These cells are moderately active in mitosis.

5. The *stratum basale* (stratum germinativum) contains a single layer of columnar cells actively dividing by mitosis to produce cells that migrate into the upper epidermal layers and ultimately to the surface of the skin.

The Dermis

The second layer of the skin, the dermis, consists of various connective tissues. As connective tissue, it contains fibroblasts and macrophages within a gelatinous matrix containing collagen, elastic, and reticular fibers. The structure provides strength, extensibility (the ability to be stretched), and elasticity (the ability to return to its original form). It is in the dermis where we find capillaries and many nerve endings. Major blood vessels are found in the hypodermis (see below).

The dermis consists of two layers:

- The *papillary layer* is a thin outer layer of areolar connective tissue with fingerlike projections called *dermal papillae* that protrude into the epidermis. In the hands and feet, the dermal papillae generate epidermal ridges (sweat from the epidermal ridges leaves fingerprints).

- The *reticular layer* is a thick layer of dense irregular connective tissue. It lies deep to the papillary layer and makes up most of the dermis.

The Hypodermis

The *hypodermis* (subcutaneous layer, or superficial fascia) lies between the dermis and underlying tissues and organs. It consists of mostly adipose tissue and is the storage site of most body fat. It serves to fasten the skin to the underlying surface, provides thermal insulation, and absorbs shocks from impacts to the skin.

Accessory Organs of the Skin

The following accessory organs (skin derivatives) are embedded in the skin:

- Hairs are elongated filaments of keratinized epithelial cells that arise and emerge from the skin of mammals. Hair is composed of the following structures:

  - The *hair shaft* is the portion of the hair that is visible on the surface of the skin.

  - The *hair root* is the portion of the hair that penetrates the skin (epidermis and dermis).

  - The *hair follicle* is the sheath that surrounds the hair in the skin.

  - The *bulb* is the base of the hair follicle.

  - The *matrix* is the bottom of the hair follicle (located within the bulb). Here, cells are actively dividing, producing new hair cells. As these cells differentiate, they produce keratin and absorb melanin from nearby melanocytes. As younger cells are produced below them, the more mature cells are pushed upward, where they eventually die. The keratin they leave behind contributes to the growth of the hair. The color of the hair is determined by the pigments absorbed from the melanocytes.

  - The *arrector pili* is a smooth muscle that is attached to the hair follicle. When the muscle contracts, the hair becomes erect; in humans, "goose bumps" are produced.

- Nails are keratinized epithelial cells. The semilunar lighter region of the nail, the lunula, is the area of new nail growth. Below the lunula, the nail matrix is actively producing nail cells, which contribute to the growth of the nail.

- Sudoriferous (sweat) glands secrete sweat. Sweat consists of water with various salts and other substances. There are four kinds of sudoriferous glands:

■ Merocrine glands occur under most skin surfaces and secrete a watery solution through pores (openings at the skin surface), which serve to cool the skin as it evaporates.

■ Apocrine glands occur under skin surfaces of the armpits and pubic regions and, beginning with puberty, secrete a solution in response to stress or sexual excitement. The solution, more viscous and more odorous than that secreted by eccrine glands, is secreted into hair follicles.

■ Ceruminous glands secrete cerumen (earwax) into the external ear canal. Wax helps to impede the entrance of foreign bodies.

■ Mammary glands produce milk that is secreted through the nipples of the breasts.

■ Sebaceous (oil) glands secrete sebum, an oily substance, into hair follicles or sometimes through skin surface pores. Sebum inhibits bacterial growth and helps prevent drying of hair and skin. An accumulation of sebum in the duct of a sebaceous gland produces whiteheads, blackheads (if the sebum oxidizes), and acne (if the sebum becomes infected by bacteria).

Chapter Check-Out

Q&A

1. Skin is very important in the synthesis of what important nutrient?
2. True or False: The dermis is the storage site of most body fat and plays an important role in thermal regulation.
3. How many layers of epidermal cells are found in the skin of our cheeks?
   a. 2
   b. 3
   c. 4
   d. 5
   e. 6

Answers: 1. vitamin D, **2.** F, **3.** c

Chapter 5

BONES AND SKELETAL TISSUES

Chapter Check-In

❏ Comprehending the various functions and types of bone

❏ Understanding the difference between compact and spongy bone

❏ Listing the main features of long, short, irregular, and flat bones

❏ Appreciating the basics of how bone develops, grows, and remodels

Bone is often stereotyped as simply a protective and supportive framework for the body. Though it does perform these functions, bone is actually a very dynamic organ that is constantly remodeling and changing shape to adapt to the daily forces placed upon it. Moreover, bone stores crucial nutrients, minerals, and lipids and produces blood cells that nourish the body and play a vital role in protecting the body against infection. All these functions make the approximately 206 bones of the human body an organ that is essential to our daily existence.

Functions of Bones

The skeletal system consists of bones, cartilage, and the membranes that line the bones. Each bone is an organ that includes nervous tissue, epithelial tissue (within the blood vessels), and connective tissue (blood, bone, cartilage, adipose, and fibrous connective tissue).

Bones have many functions, including the following:

- *Support:* Bones provide a framework for the attachment of muscles and other tissues.

- *Protection:* Bones such as the skull and rib cage protect internal organs from injury.

- *Movement:* Bones enable body movements by acting as levers and points of attachment for muscles.

- *Mineral storage:* Bones serve as a reservoir for calcium and phosphorus, essential minerals for various cellular activities throughout the body.

- *Blood cell production:* The production of blood cells, or hematopoiesis, occurs in the red marrow found within the cavities of certain bones.

- *Energy storage:* Lipids, such as fats, stored in adipose cells of the yellow marrow serve as an energy reservoir.

Types of Bones

Bones come in several different types. Long bones are longer than they are wide. The length of the bone, or shaft, widens at the extremities (ends). Short bones are cubelike, about as long as they are wide. Flat bones, such as ribs or skull bones, are thin or flattened. Irregular bones, such as vertebrae, facial bones, or hip bones, have specific shapes, unlike the other types of bones.

The following two bone types are usually classified separately:

- Sesamoid or round bones, such as the kneecap, are found embedded within certain tendons.

- Sutural or **Wormian bones** occur between the sutures (joints) of the cranial bones of the skull.

Bone Structure

There are two kinds of bone tissue (see Figure 5-1):

Figure 5-1 Main features of a long bone.

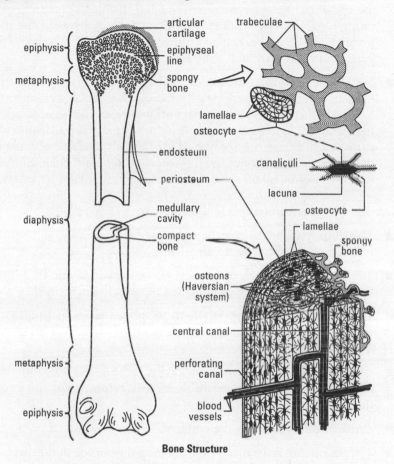

Bone Structure

■ Compact bone is the hard material that makes up the shaft of long bones and the outside surfaces of other bones. Compact bone consists of cylindrical units called osteons. Each osteon contains concentric lamellae (layers) of hard, calcified matrix with osteocytes (bone cells) lodged in lacunae (spaces) between the lamellae. Smaller canals, or canaliculi, radiate outward from a central canal, which contains blood vessels and nerve fibers. Osteocytes within an osteon are connected to each other and to the central canal by fine cellular extensions. Through these cellular extensions, nutrients and waste are exchanged between the osteocytes and the blood vessels. Perforating canals provide channels that allow the blood vessels that run through the central canals to connect to the blood vessels in the periosteum that surrounds the bone.

■ Spongy bone consists of thin, irregularly shaped plates called trabeculae, arranged in a latticework network. Trabeculae are similar to osteons in that both have osteocytes in lacunae that lie between calcified lamellae. As in osteons, canaliculi present in trabeculae provide connections between osteocytes. However, since each trabecula is only a few cell layers thick, each osteocyte is able to exchange nutrients with nearby blood vessels. Thus, no central canal is necessary.

Here are the main features of a long bone (refer to Figure 5-1):

■ The **diaphysis,** or shaft, is the long tubular portion of long bones. It is composed of compact bone tissue.

■ The **epiphysis** (plural, epiphyses) is the expanded end of a long bone. It is in the epiphyses where red blood cells are formed.

■ The metaphysis is the area where the diaphysis meets the epiphysis. It includes the epiphyseal line, a remnant of cartilage from growing bones.

■ The medullary cavity, or marrow cavity, is the open area within the diaphysis. The adipose tissue inside the cavity stores lipids and forms the yellow marrow.

■ Articular cartilage covers the epiphysis where joints occur.

■ The periosteum is the membrane covering the outside of the diaphysis (and epiphyses where articular cartilage is absent). It contains osteoblasts (bone-forming cells), osteoclasts (bone-destroying cells), nerve fibers, and blood and lymphatic vessels. Ligaments and tendons attach to the periosteum.

■ The endosteum is the membrane that lines the marrow cavity.

Here are the main features of short, flat, and irregular bones:

- In short and irregular bones, spongy bone tissue is encircled by a thin layer of compact bone tissue.

- In flat bones, the spongy bone tissue is sandwiched between two layers of compact bone tissue. The spongy bone tissue is called the diploë.

- The periosteum covers the outside layer of compact bone tissue.

- The endosteum covers the trabeculae that fill the inside of the bone.

- In certain bones (ribs, vertebrae, hip bones, sternum), the spaces between the trabeculae contain red marrow, which is active in hematopoiesis.

Bone Development

The skeleton arises from fibrous membranes and hyaline cartilage during the first month of embryonic development. These tissues are replaced with bone by two different bone-building, or ossification, processes.

The first process, called intramembranous ossification, occurs when fibrous membranes are replaced by bone tissue. The process, occurring only in certain flat bones, such as the flat bones of the skull, sternum, and clavicle, is summarized in two basic steps:

1. Spongy bone tissue begins to develop at sites within the membranes called centers of ossification.

2. Red bone marrow forms within the spongy bone tissue, followed by the formation of compact bone on the outside.

The second ossification process, called endochondral ossification, occurs when hyaline cartilage is replaced by bone tissue. The process, occurring in most bones of the body, follows these steps:

1. At a primary ossification center, in the center of a cartilage model, hyaline cartilage breaks down, forming a cavity.

2. A periosteal bud, consisting of osteoblasts, osteoclasts, red marrow, nerves, and blood and lymph vessels, invades the cavity. The osteoblasts produce spongy bone tissue.

3. A medullary cavity forms as osteoclasts break down the newly produced spongy bone tissue. The medullary cavity expands as it follows the spread of the primary ossification center to the ends of the bone.

4. Compact bone tissue replaces cartilage on the outside of the bone.

5. In long bones, secondary ossification centers form in the epiphyses. As in the shaft, a periosteal bud develops. However, the spongy bone tissue that subsequently develops is not replaced by a medullary cavity.

6. Articular cartilage is formed from cartilage remaining on the outside of the epiphyses.

7. The epiphyseal plate is formed from cartilage remaining between the expanding primary and secondary ossification centers.

Bone Growth

Bones elongate as chondrocytes in the cartilage of the epiphyseal plate divide. These cell divisions produce new cartilage within the epiphyseal plate bordering the epiphyses. At the other end of the epiphyseal plate, bordering the diaphysis, older cartilage is broken down by invading osteoclasts and eventually replaced by the expanding medullary cavity.

Bone Homeostasis

Remodeling is the process of creating new bone and removing old bone. It occurs constantly in growing children as well as in adults in the following situations:

- When bones grow, remodeling causes bone tissue to be redistributed to maintain the shape and structure of the bone.

- In response to new stresses applied to a bone, remodeling increases bone strength by adding new bone tissue where appropriate.

- Remodeling enables calcium stored in the bones to be removed for metabolic processes in other parts of the body. Similarly, remodeling occurs when excess calcium is returned to the bone reservoir.

- Remodeling occurs during the repair of broken bones.

Surface Features of Bones

The surfaces of bones bear projections, depressions, ridges, and various other features. A process (projection) on one bone may fit with a depression on a second bone to form a joint. Another process allows for the attachment of a muscle or ligament. Grooves and openings provide passageways for blood vessels or nerves. A list of the various processes and other surface features appears in Table 5-1.

Table 5-1 Various Processes and Other Surface Features

General	Process	Projection or Prominence on a Bone
Processes that help form joints	Condyle	Large, rounded articular process
	Facet	Smooth, flat surface
	Head	Enlarged portion at an end of a bone
	Ramus	Branch or extension of a bone

(continued)

Table 5-1 *(continued)*

General	Process	Projection or Prominence on a Bone
Processes that provide for the attachment of muscles and ligaments	Crest	Narrow ridge
	Epicondyle	Process on or above a condyle
	Linea (line)	Narrow ridge (less prominent than a crest)
	Spine	Sharp or pointed process (spinous process)
	Trochanter	Large, irregularly shaped process (found only on the femur)
	Tubercle	Small, knoblike process
	Tuberosity	Large, knoblike process
Depressions or openings (may provide passageways for blood vessels and nerves)	Fissure	Narrow opening
	Fontanel	Membrane-covered spaces between skull bones
	Foramen	Round opening
	Fossa	Shallow depression
	Fovea	Pitlike depression
	Meatus	Tubelike passage
	Sinus	Interior cavity
	Sulcus (or groove)	Long, narrow depression

Chapter Check-Out

Q&A

1. In spongy bone, there is a meshwork of thin, irregularly shaped plates called _____.

2. True or False: Some bones are completely embedded in tendons.

3. Which of the following bone (or bones) represent(s) a flat bone?
   a. the frontal bone
   b. the humerus
   c. the ninth rib
   d. a cervical vertebra

4. True or False: The skeleton arises from fibrous membranes and elastic cartilage during the early stages of embryonic development.

5. The _____ is a remnant of cartilage found in growing long bones.

Answers: 1. trabeculae, **2.** T (sesamoid bones), **3.** a and c, **4.** F, **5.** epiphyseal line

Chapter 6

THE SKELETAL SYSTEM

Chapter Check-In

❑ Understanding the differences between the axial and appendicular
skeleton

❑ Identifying the bony composition of each part of the skeleton

❑ Listing the names of the various bones of the body

Knowing the skeletal organization of each part of the body is crucial
in anatomy. Though memorizing the names and locations of all
206 bones may be a tedious process, this knowledge can be very valu-
able. This is especially true if you are planning a future in science, want
to understand the language your physician is using, hope to do well in
an anatomy course, or peruse scientific literature.

Organization of the Skeleton

The bones of the body are categorized into two groups: the axial skeleton
and the appendicular skeleton. The bones of the axial skeleton revolve
around the vertical axis of the skeleton, while the bones of the appendicular skeleton make up the limbs that have been appended to the axial
skeleton. A list of the bones in the axial and appendicular skeletons is
given in Tables 6-1 and 6-2, respectively. Major bones of both skeletons
are shown in Figure 6-1.

Figure 6-1 Major bones of the axial and appendicular skeletons.

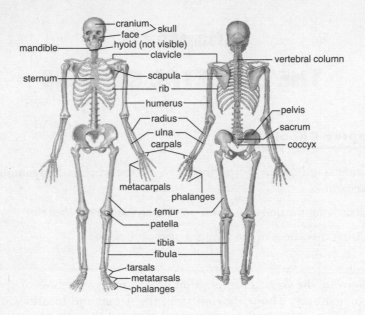

Table 6-1   Axial Skeleton Bones (80 Bones)

General Description	Name of Bone	No. of Bones	Additional Information
cranium (8)	frontal	1	
	parietal	2	
	temporal	2	
	sphenoid	1	
	ethmoid	1	
	occipital	1	
facial bones (14)	mandible	1	lower jawbone
	maxilla	2	upper jawbone
	zygomatic bone	2	cheek bones
	nasal bone	2	
	lacrimal bone	2	
	palatine	2	
	inferior nasal concha	2	
	vomer	1	
hyoid	hyoid	1	
ear ossicles	malleus, incus, stapes	6	2 each
vertebral column (26)	cervical vertebrae	7	C_1–C_7
	thoracic vertebrae	12	T_1–T_{12}
	lumbar vertebrae	5	L_1–L_5
	sacrum	1	5 fused S_1–S_5
	coccyx	1	3–5 fused
thorax (thoracic cage or bony thorax)	sternum	1	
	true ribs (7 pair)	14	vertebrosternal
	false ribs (3 pair)	6	vertebrochondral
	false ribs (floating [2 pair] ribs)	4	vertebral ribs

Table 6-2   Appendicular Skeleton Bones (126 Bones)

General Description	Name of Bone	No. of Bones	Additional Information
pectoral girdle (shoulder girdle)	clavicle scapula	2 2	collarbone shoulder blade
upper limb (60)	humerus	2	1 per upper arm
	ulna	2	1 per forearm
	radius	2	8 per wrist
	carpals (16)		
	1. scaphoid	2	
	2. trapezium	2	
	3. capitate	2	
	4. trapezoid	2	
	5. lunate	2	
	6. triquetrum	2	
	7. pisiform	2	
	8. hamate	2	
	metacarpals	10	5 in each hand
	phalanges	28	3 per digit (pollex has 2)
pelvic girdle (hip) (2)	coxal bones (os coxae)	2	hip bones
	1. ilium		
	2. ischium		(3 fused pairs)
	3. pubis		
lower limb (60)	femur	2	1 per upper leg
	patella	2	1 per leg
	tibia	2	1 per lower leg
	fibula	2	1 per lower leg
	tarsal (14)		7 per ankle
	1. talus	2	
	2. calcaneus	2	
	3. cuboid	2	
	4. navicular	2	
	5. medial cuneiform	2	
	6. intermediate cuneiform	2	
	7. lateral cuneiform	2	
	metatarsals	10	5 per foot
	phalanges	28	3 per digit (hallux has 2)

Skull: Cranium and Facial Bones

The skull consists of 8 cranial bones and 14 facial bones. The bones are listed in Table 6-1, but note that only six types of cranial bones and eight types of facial bones are listed because some of the bones (as indicated in the table) exist as pairs.

The bones of the skull provide protection for the brain and the organs of vision, taste, hearing, equilibrium, and smell. The bones also provide attachment for muscles that move the head and control facial expressions and chewing.

Figures 6-2 and 6-3 illustrate specific characteristics of these bones, while some general features of the skull follow:

■ Sutures are immovable interlocking joints that join skull bones together.

■ Fontanels are spaces between cranial bones that are filled with fibrous membranes. The spaces provide pliability for the skull when it passes through the birth canal and for brain growth during infancy. Bone growth eventually fills the spaces by age two.

■ Sutural (Wormian) bones are very small bones that develop within sutures. Their number and location vary.

■ The cranial vault denotes the top, sides, front, and back of the cranium. The cranial floor (base) denotes the bottom of the cranium.

■ Cranial fossae are three depressions in the floor of the cranium. These fossae, called the anterior, middle, and posterior cranial fossae, provide spaces that accommodate the shape of the brain.

■ The nasal cavity is formed by cartilage and several bones. Air entering the cavity is warmed and cleansed by mucus lining the cavity.

■ Sinuses (paranasal sinuses) are mucus-lined cavities inside cranial and facial bones that surround the nasal cavity. The cavities secrete mucus that drains into the nasal cavity. The cavities also act as resonance chambers that enhance vocal (and singing) quality.

Figure 6-2 The right lateral view and anterior view of the skull's bones

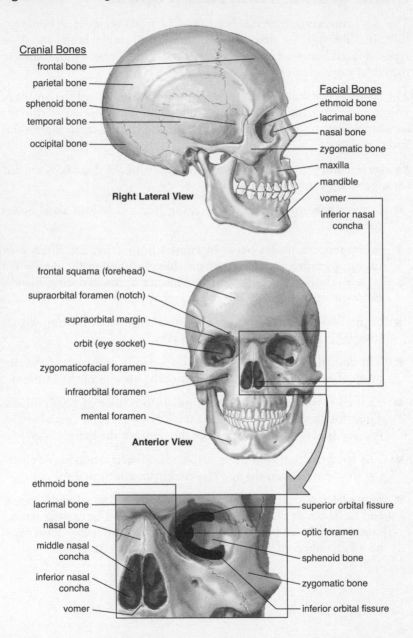

Cranial Bones
frontal bone
parietal bone
sphenoid bone
temporal bone
occipital bone

Facial Bones
ethmoid bone
lacrimal bone
nasal bone
zygomatic bone
maxilla
mandible
vomer
inferior nasal concha

Right Lateral View

frontal squama (forehead)
supraorbital foramen (notch)
supraorbital margin
orbit (eye socket)
zygomaticofacial foramen
infraorbital foramen
mental foramen

Anterior View

ethmoid bone
lacrimal bone
nasal bone
middle nasal concha
inferior nasal concha
vomer

superior orbital fissure
optic foramen
sphenoid bone
zygomatic bone
inferior orbital fissure

Figure 6-3   The sagittal section and interior view of the skull's bones

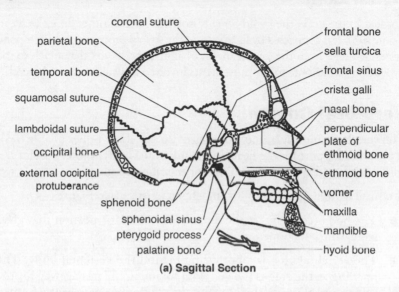

(a) Sagittal Section

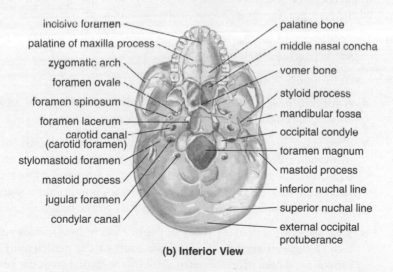

(b) Inferior View

Hyoid Bone

Located in the neck, the hyoid bone is isolated from all other bones (refer to Figure 6-3). It is connected by ligaments to the styloid processes of the temporal bones. Muscles from the tongue, neck, pharynx, and larynx that attach to the hyoid bone contribute to the movements involved in swallowing and speech.

Vertebral Column

The vertebral column (spine) consists of 26 vertebrae bones (Table 6-3). It provides support for the head and trunk of the body, protection for the spinal cord, and connecting points for the ribs and muscles.

A typical vertebra has the following characteristics (see Figure 6-4):

- The body (centrum) is the disc-shaped, anterior portion that gives strength to the bone.

- The vertebral arch is a bony ring behind the vertebral body. The opening in the ring is the vertebral foramen, the passageway for the spinal cord. The vertebral canal is the continuous passageway formed by the vertebral foramina of successive vertebrae.

- The pedicles and the laminae form the anterior and posterior sides, respectively, of the vertebral arch.

- Seven processes project from the vertebral arch:

  - A spinous process projects posteriorly from the vertebral arch. Muscle and ligaments attach to the spinous process.

  - Two transverse processes project from the vertebral arch, one from each side, at each of the junctures of the pedicles and laminae. Muscles and ligaments attach to the transverse processes. Each transverse process of cervical vertebrae contains a transverse foramen through which blood vessels pass to the brain.

  - Two superior articular processes project from the superior surface of the vertebral arch, one from each of the pediclelamina junctions. These processes articulate (form joints) with the preceding vertebrae.

  - Two inferior articular processes project from the inferior surface of the vertebral arch, one from each of the pediclelamina junctions. These processes articulate (form joints) with the vertebra next in line.

Table 6-3   Bones of the Vertebral Column

Vertebra	Body	Spinous Process	Transverse Process	Vertebral Opening
Cervical				
C_1 (atlas)	none, bony ring	none	with transverse foramen	lightbulb-shaped
C_2 (axis)	relatively small; with dens (odontoid process)	bifid	with transverse foramen	large; heart-shaped
C_3-C_6	relatively small; oval	bifid	with transverse foramen	large; triangular
C_7 (vertebra prominens)	relatively small; oval	prominently long; not bifid	with transverse foramen	large; triangular
Thoracic				
T_1-T_{10}	larger than C_2-C_7, heart-shaped; 2 facets or demifacets for articulating with rib head	long; points inferiorly	with facets for articulating with rib tubercle	circular
$T_{11}-T_{12}$	larger than C_2-C_7, heart shaped; 1 demifacets for articulating with rib head	long; points inferiorly	no facets for rib joints	circular
Lumbar				
L_1-L_5	largest of all vertebrae	short, thick; points horizontally		
Sacrum				
5 fused S_1-S_5	fusion of 5 vertebrae forms a triangular bone	fuse to become the median sacral crest	becomes the lateral sacral crest	becomes sacral canal
Coccyx				
4 fused				

Figure 6-4 The four regions of the vertebral column.

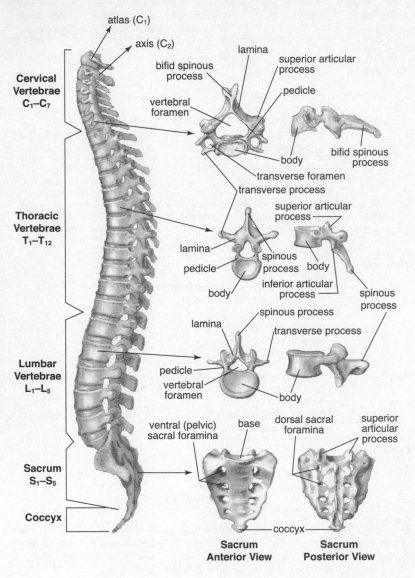

Cervical Vertebrae C₁–C₇

Thoracic Vertebrae T₁–T₁₂

Lumbar Vertebrae L₁–L₅

Sacrum S₁–S₅

Coccyx

atlas (C₁)

axis (C₂)

lamina

bifid spinous process

superior articular process

pedicle

vertebral foramen

bifid spinous process

body

transverse foramen

transverse process

superior articular process

lamina

pedicle

spinous process

body

body

inferior articular process

spinous process

spinous process

lamina

transverse process

pedicle

vertebral foramen

body

ventral (pelvic) sacral foramina

base

dorsal sacral foramina

superior articular process

coccyx

Sacrum Anterior View

Sacrum Posterior View

- The intervertebral foramina are openings between the superior and inferior surfaces of each pedicle of the vertebral arch. Adjacent openings of successive vertebrae form a passage for nerves that leave the spinal cord and emerge outside the vertebral column.

- Intervertebral discs separate adjacent vertebrae. Each disc consists of an outer ring of fibrocartilage (annulus fibrosus) surrounding a semi-fluid cushion (nucleus pulposus) that provides elasticity and compressibility.

The vertebral column is divided into four regions, with each region contributing to the alternating concave and convex curves of the spine (see Figure 6-4). As the vertebrae progress down the column, their bodies get more massive, enabling them to bear more weight.

The sacrum is a triangular bone below the last lumbar vertebra (see Figure 6-4). It is formed by the fusion of five vertebrae (S_1–S_5).

The coccyx, formed by four fused vertebrae, is a small triangle-shaped bone that attaches to the bottom of the sacrum (see Figure 6-4).

Thorax

The thoracic cage includes the thoracic vertebrae, sternum, ribs, and costal cartilage (see Figure 6-5). The sternum (breastbone) consists of three fused bones: the manubrium, body, and xiphoid process. There are 12 pairs of ribs. All ribs articulate posteriorly with a corresponding thoracic vertebra. At their anterior ends, they differ as to how they attach, as follows:

- Seven pairs of true ribs (vertebrosternal ribs) attach directly to the sternum with hyaline cartilage called costal cartilage.

- Three pairs of false ribs (vertebrochondral ribs) do not attach to the sternum. Rather, they connect (with costal cartilage) to the rib directly above them.

- Two pairs of false ribs (floating ribs or vertebral ribs) do not attach to anything at their anterior ends.

Figure 6-5 The thoracic cage.

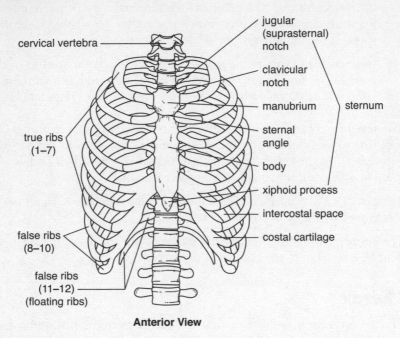

Anterior View

Here are important features of a rib:

- The head is the end of the rib that articulates with the vertebral column.

- The superior and inferior facets on the head articulate with the facets of the thoracic vertebrae.

- The neck, just beyond the head, bears a tubercle (rounded process) that articulates with the facet of the vertebral transverse process. Part of the tubercle also presents a place of attachment for ligaments.

- The costal angle designates the sharp turn of the rib.

- The costal groove, a passageway on the inside of the bending rib, provides for blood vessels and intercostal nerves.

- The body (shaft) is the major part of the rib—that part beyond the costal angle.

- Intercostal spaces, the areas between the ribs, are occupied by the intercostal muscles.

Pectoral Girdle

Each of the two pectoral (shoulder) girdles consists of two bones: the S-shaped clavicle and the flat, triangular scapula. The clavicle articulates with the sternum and the scapula. In turn, the scapula articulates with the humerus of the arm. Figure 6-6 illustrates details of these bones.

Figure 6-6   The pectoral girdle.

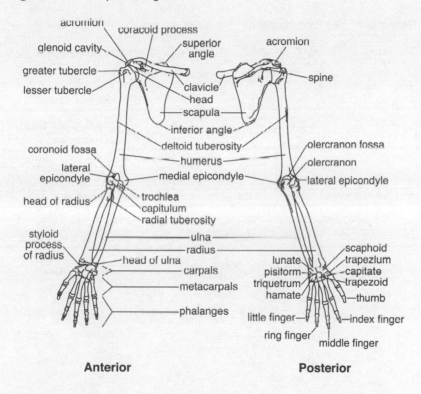

Anterior   Posterior

Upper Limb

The upper limb consists of the arm, forearm, and hand. The 30 bones of each upper limb are illustrated in Figure 6-6.

Pelvic Girdle

The pelvic (hip) girdle transfers the weight of the upper body to the legs. It consists of a pair of coxal bones (os coxae, hip bones), each of which contains three fused bones: the ilium, ischium, and pubis. Together with the sacrum and coccyx, the pelvic girdle forms a bowl-shaped region, the pelvis, that protects internal reproductive organs, the urinary bladder, and the lower part of the digestive tract. Features of these bones are given in Figure 6-7.

Figure 6-7   The pelvic girdle

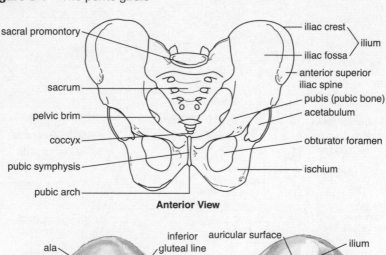

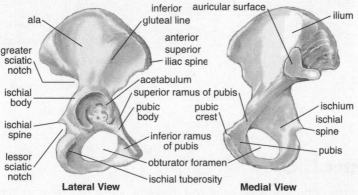

Lower Limb

The thigh, leg, and foot constitute the lower limb. The bones of the lower limbs are considerably larger and stronger than comparable bones of the upper limbs because the lower limbs must support the entire weight of the body while walking, running, or jumping. Figure 6-8 illustrates features of the 30 bones of each lower limb.

Figure 6-8 The 30 bones of each lower limb.

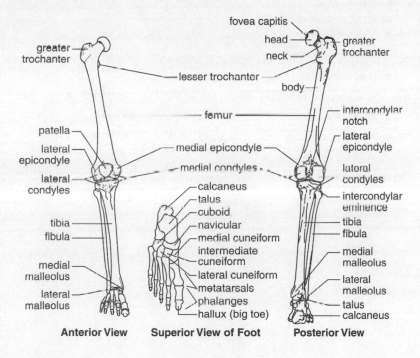

Anterior View **Superior View of Foot** **Posterior View**

Chapter Check-Out

Q&A

1. A bone crucial to swallowing and speech is the _____ _____.

2. True or False: The sphenoid and occipital bones form most of the base of the cranium.

3. Which of the following bone (or bones) help(s) compose the face?
   a. the parietal bones
   b. the talus
   c. the zygomatic bone
   d. the vomer

4. True or False: The coccyx never varies in the number of vertebrae that compose it.

5. The outer, fibrocartilaginous ring of the intervertebral disk is called the _____ _____.

Answers: 1. hyoid bone, **2.** T, **3.** c and d, **4.** F, **5.** annulus fibrosus

Chapter 7

ARTICULATIONS

Chapter Check-In

❑ Discovering the different types of joints

❑ Understanding the components of a joint

❑ Classifying joints

The intricate movements of a human, such as those performed in dance and athletics, are accomplished by using a wide variety of joints. Though joints enable the skeleton to be dynamic, they also play an important role in stability and protection. In fact, the mobility of a joint is often inversely proportional to its stability. For example, the sutures of the bones of the cranium are basically immovable in relationship to one another, but due to their stable nature, they serve to protect the brain throughout daily life and during incidents of trauma. On the other hand, the ball-and-socket of the shoulder enables a wide variety of complex movements. This increase in the amount of mobility leads to instability, which is why the shoulder is susceptible to injury. The goal of this chapter is to familiarize you with the wide variety of joints that are in the human body and to help you understand how these joints are classified.

Classifying Joints

A joint (articulation) occurs wherever bones meet. Joints are classified both structurally and functionally, as shown in Table 7-1.

Table 7-1 Joint Classification

Functional Class	Structural Class	Joint	Description Type	Example of Joint
synarthrosis (immovable)	fibrous	suture	interlocking seams	between cranial bones
synarthrosis (immovable)	fibrous	gomphosis	peg-and-socket joint	between teeth and sockets
synarthrosis (immovable)	cartilaginous	synchrondrosis	hyaline cartilage joint	between diaphysis and epiphysis in long bones
amphiarthrosis (slightly movable)	fibrous	syndesmosis	ligament or distal interosseous membrane	joint of tibia and fibula
amphiarthrosis (slightly movable)	cartilaginous	symphysis	fibrocartilage acts as compressible cushion	intervertebral discs of vertebral column
diarthrosis (freely movable)	synovial	gliding	two sliding surfaces	between carpals
diarthrosis (freely movable)	synovial	hinge	concave surface with convex surface	between humerus and ulna
diarthrosis (freely movable)	synovial	pivot	rounded end fits into ring of bone and ligament	between altas (C_1) and axis (C_2) vertebrae
diarthrosis (freely movable)	synovial	condyloid	oval condyle with oval cavity	between metacarpals and phalanges
diarthrosis (freely movable)	synovial	saddle	each surface is both concave and convex	between carpus and the first metacarpal
diarthrosis (freely movable)	synovial	ball-and-socket	ball-shaped head with cup-shaped socket	between femur and pelvis

Structural classification

Structural classification is based on the materials that hold the joint together and whether or not a cavity is present in the joint. There are three structural classes:

- Fibrous joints are held together by fibrous connective tissue. No joint cavity is present. Fibrous joints may be immovable or slightly movable.

- Cartilaginous joints are held together by cartilage (hyaline or fibrocartilage). No joint cavity is present. Cartilaginous joints may be immovable or slightly movable.

- **Synovial joints** are characterized by a synovial cavity (joint cavity) containing synovial fluid. Synovial joints are freely movable and characterize most joints of the body. Figure 7-1 lists other features of a synovial joint, including the following:

  - Articular cartilage (hyaline cartilage), which covers the end of each bone.

  - A synovial membrane, which surrounds the synovial cavity. Its areolar connective tissue secretes a lubricating synovial fluid into the synovial cavity.

  - A fibrous capsule outside the synovial membrane, which surrounds the joint. It often contains bundles of dense, irregular, connective tissue called ligaments. The ligaments provide strength and flexibility to the joint.

  - The articulate capsule, composed of the synovial membrane and fibrous capsule.

  - Accessory ligaments, which lie outside the articular capsule (extracapsular ligaments) or inside the synovial cavity (intracapsular ligaments).

Figure 7-1   A synovial joint.

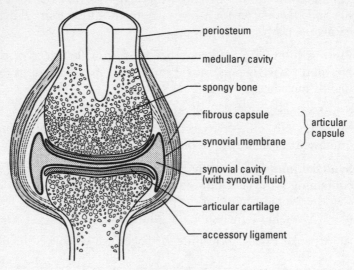

periosteum

medullary cavity

spongy bone

fibrous capsule

synovial membrane

} articular capsule

synovial cavity
(with synovial fluid)

articular cartilage

accessory ligament

A Synovial Joint

Functional classification

Functional classification is based on the degree to which the joint permits movement. There are three types:

- A **synarthrosis** joint permits no movement. Structurally, it may be a fibrous or cartilaginous joint.

- An **amphiarthrosis** joint permits only slight movement. Structurally, it may be a fibrous or cartilaginous joint.

- A **diarthrosis** joint is a freely movable joint. Structurally, it is always a synovial joint.

Chapter Check-Out

Q&A

1. Name the two types of joints where no joint cavity is present.

2. True or False: A diarthrosis joint allows little or no movement.

3. Which of the following is an example of an amphiarthrotic joint?

   a. between the femur and pelvis
   b. between the teeth and their sockets
   c. between cranial bones
   d. between the intervertebral disks
   e. between the metacarpals and phalanges

Answers: 1. fibrous and cartilaginous joints, **2.** F, **3.** d

Chapter 8

MUSCLE TISSUE

Chapter Check-In

❑ Discussing the structure and function of different muscle types

❑ Understanding the connective tissue associated with skeletal muscle

❑ Describing the physiology and stimulation of a muscular contraction

❑ Appreciating major metabolic pathways involved in cellular respiration

Muscle enables complex movements that are either voluntary—under conscious control—such as turning the pages of this book, or involuntary, such as the contraction of the heart or the peristalsis in the gut. To understand how muscle accomplishes these various activities, you need to know the physiology behind a muscle contraction. This requires a detailed knowledge of the muscle's microscopic anatomy. Of course, muscle contractions will not take place without adequate nervous stimulation or a sufficient supply of ATP, the muscles' fuel. ATP is obtained via cellular respiration, which is accomplished by several different metabolic pathways discussed in this chapter.

Types of Muscles

There are three types of muscles:

- *Skeletal muscles* are attached mainly to the skeletal bones but some are also attached to other structures (such as the eyes for eye movement) and causes movements of the body. Skeletal muscle is also called striated muscle, because of its banding pattern when viewed under a microscope (for clarification, see cardiac muscle below), or voluntary muscle (because muscle contractions can be consciously controlled).

- *Cardiac muscle* is responsible for the rhythmic contractions of the heart. Cardiac muscle is involuntary—it generates its own stimuli to initiate a muscle contraction. While cardiac muscle also consists of striations, the main characteristic (to differentiate these striations from skeletal muscle) is the presence of intercalated disks.

- *Smooth muscle* lines the walls of hollow organs. For example, it lines the walls of blood vessels and of the digestive tract, where it serves to advance the movement of substances. A smooth muscle contraction is relatively slow and involuntary.

Connective Tissue Associated with Muscle Tissue

A skeletal muscle consists of numerous muscle cells called muscle fibers. Three layers of connective tissues surround these fibers to form a muscle. These and other connective tissues associated with muscles follow:

- The *endomysium* is the connective tissue that surrounds each muscle fiber (cell).

- The *perimysium* encircles a group of muscle fibers, forming a fascicle.

- The *epimysium* encircles all the fascicles to form a complete muscle.

- A *tendon* is a cordlike extension of the preceding three linings. It extends beyond the muscle tissue to connect the muscle to a bone or to other muscles.

- An **aponeurosis** is a flat broad extension of the three muscle linings and serves the same function as a tendon.

- **Fascia** is a term for a layer or sheet of connective tissue.

  - The *deep fascia* surrounds the epimysium and encloses or lines other nearby structures such as blood vessels, nerves, and the body wall.

  - The *superficial fascia* (hypodermis or subcutaneous layer) lies immediately below the skin. The superficial fascia merges with the deep fascia where the surfaces of the skin meet.

Structure of Skeletal Muscle

A muscle fiber (cell) has special terminology and distinguishing characteristics:

- The **sarcolemma,** or plasma membrane of the muscle cell, is highly invaginated by transverse tubules (T tubes) that permeate the cell.

- The sarcoplasm, or cytoplasm of the muscle cell, contains calcium-storing sarcoplasmic reticulum, the specialized endoplasmic reticulum of a muscle cell.

- Striated muscle cells are multinucleated. The nuclei lie along the periphery of the cell, forming swellings visible through the sarcolemma.

- Nearly the entire volume of the cell is filled with numerous long myofibrils. Myofibrils consist of two types of filaments, shown in Figure 8-1:

  - Thin filaments consist of two strands of the globular protein actin arranged in a double helix. Along the length of the helix are troponin and tropomyosin molecules that cover special binding sites on the actin.

  - Thick filaments consist of groups of the filamentous protein myosin. Each myosin filament forms a protruding head at one end. An array of myosin filaments possesses protruding heads at numerous positions at both ends.

Figure 8-1 Two types of filaments.

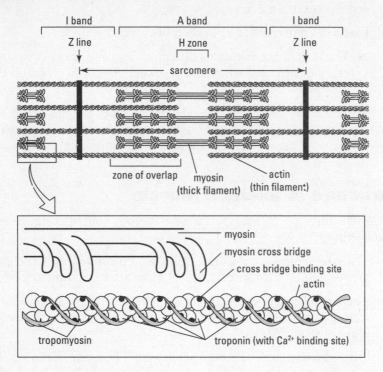

Within a myofibril, **actin** and myosin filaments are parallel and arranged side by side. The overlapping filaments produce a repeating pattern that gives skeletal muscle its striated appearance. Each repeating unit of the pattern, called a **sarcomere,** is separated by a border, or Z disc (Z line), to which the actin filaments are attached. The myosin filaments, with their protruding heads, float between the actin, unattached to the Z disc.

Muscle Contraction

Muscle contraction events describing the sliding-filament concept are listed as follows.

1. ATP binds to a myosin head and forms ADP + P_i. When ATP binds to a myosin head, it is converted to ADP and P_i, which remain attached to the myosin head.

2. Ca^{2+} exposes the binding sites on the actin filaments. Ca^{2+} binds to the troponin molecule, causing tropomyosin to expose positions on the actin filament for the attachment of myosin heads.

3. When attachment sites on the actin are exposed, the myosin heads bind to actin to form cross bridges.

4. ADP and P_i are released, and a sliding motion of actin results. The attachment of cross bridges between myosin and actin causes the release of ADP and P_i. This, in turn, causes a change in the shape of the myosin head, which generates a sliding movement of the actin toward the center of the sarcomere. This pulls the two Z discs together, effectively contracting the muscle fiber to produce a power stroke.

5. ATP causes the cross bridges to unbind. When a new ATP molecule attaches to the myosin head, the cross bridge between the actin and myosin breaks, returning the myosin head to its unattached position.

Without the addition of a new ATP molecule, the cross bridges remain attached to the actin filaments. This is why corpses become stiff with rigor mortis (new ATP molecules are unavailable).

Stimulation of muscle contraction

Neurons, or nerve cells, are stimulated when the polarity across their plasma membrane changes. The polarity change, called an **action potential,** travels along the neuron until it reaches the end of the neuron. A gap called a **synapse** or synaptic cleft separates the neuron from a muscle cell or another neuron. If a neuron stimulates a muscle, then the neuron is a **motor neuron,** and its specialized synapse is called a neuromuscular junction. Muscle contraction is stimulated through the following steps:

1. Action potential generates release of **acetylcholine.** When an action potential of a neuron reaches the neuromuscular junction, the neuron secretes the neurotransmitter acetylcholine (Ach), which diffuses across the synaptic cleft.

2. Action potential is generated on the motor end plate and throughout the T tubules. Receptors on the motor end plate, a highly folded region of the sarcolemma, initiate an action potential. The action potential travels along the sarcolemma throughout the transverse system of tubules.

3. Sarcoplasmic reticulum releases Ca^{2+}. As a result of the action potential throughout the transverse system of tubules, the sarcoplasmic reticulum releases Ca^{2+}.

4. Myosin cross bridges form. The Ca^{2+} released by the sarcoplasmic reticulum binds to troponin molecules on the actin helix, prompting tropomyosin molecules to expose binding sites for myosin cross-bridge formation. If ATP is available, muscle contraction begins.

Phases of a muscle contraction

A muscle contraction in response to a single nerve action potential is called a twitch contraction. A myogram, a graph of muscle strength (tension) with time, shows several phases, shown in Figure 8-2:

1. The *latent period* is the time required for the release of Ca^{2+}.

2. The *contraction period* represents the time during actual muscle contraction.

3. The *relaxation period* is the time during which Ca^{2+} are returned to the sarcoplasmic reticulum by active transport.

4. The *refractory period* is the time immediately following a stimulus. This is the time period when a muscle is contracting and therefore will not respond to a second stimulus. Since this is occurring at the same time as the contraction, it does not appear on the myogram as a separate event.

Figure 8-2 The phases of a myogram.

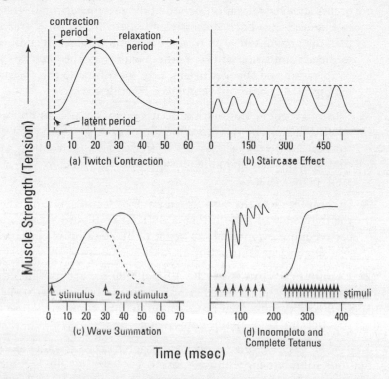

Quality of a muscle contraction

The following factors contribute to the strength and maximum duration of a muscle contraction:

- *Frequency of stimuli.* If stimuli are repeatedly applied to a muscle fiber, Ca^{2+} may not be completely transported back into the sarcoplasmic reticulum before the next stimulus occurs. Depending upon the frequency of stimuli, Ca^{2+} may accumulate. In turn, the extra Ca^{2+} results in more power strokes and a stronger muscle contraction. Depending upon the frequency of stimuli, several effects are observed:

- A staircase effect (treppe) is produced if each successive stimulus occurs after the relaxation period of the previous stimulus (refer to Figure 8-2b). Each successive muscle contraction is greater than the previous one, up to some maximum value. In addition to the accumulation of Ca^{2+}, other factors, such as increases in temperature and changes in pH, may contribute to this "warming up" effect commonly employed by athletes.

- Wave (temporal) summation occurs if consecutive stimuli are applied during the relaxation period of each preceding muscle contraction (refer to Figure 8-2c). In this case, each subsequent contraction builds upon the previous contraction before its relaxation period ends.

- **Incomplete tetanus,** also called unfused tetanus, occurs when the frequency of stimuli increases (refer to Figure 8-2d). Successive muscle contractions begin to blend, almost appearing as a single large contraction.

- **Complete tetanus,** also called fused tetanus, occurs when the frequency of stimuli increases still further (refer to Figure 8-2d). In this case, individual muscle contractions completely fuse to produce one large muscle contraction.

- *Strength of stimulus.* Muscle contractions intensify when more motor neurons stimulate more muscle fibers. This effect, called recruitment or multiple motor unit summation, is also responsible for fine motor coordination, because by continually varying the stimulation of specific muscle fibers, smooth body movements are maintained.

- *Length of muscle fiber contraction.* Because a muscle is attached to bones, muscle contraction is restricted to lengths that are between 60 percent and 175 percent of the length that produces optimal strength. This range of muscle lengths limits myosin cross bridges and actin to positions only where they overlap and thus can generate contractions.

- *Type of contraction.* Muscle contraction implies that movement occurs between myosin cross bridges and actin. However, this movement does not necessarily result in shortening of the muscle. As a result, two kinds of muscle contractions are defined:

  - Isotonic contractions occur when muscles change length during a contraction. Picking up a book is an example.

  - Isometric contractions occur when muscles do not change length during a contraction. When holding a book in midair, muscle fibers produce a force, but no motion is generated.

- *Type of muscle fiber.* Muscle fibers are classified into two groups:

  - Slow fibers contract slowly, have a high endurance, and are red from their rich blood supply. However, they do not produce much strength. These fibers are used for long-distance running.

  - Fast fibers contract rapidly, fatigue rapidly, and are white because the blood supply is limited. They generate considerable strength. These fibers are used for short-distance running.

- *Muscle tone.* In any relaxed skeletal muscle, a small number of contractions continuously occur. Observed as firmness in a muscle, these contractions maintain body posture and increase muscle readiness.

- *Muscle fatigue.* Muscle fibers stop contraction when inadequate amounts of ATP are available. Lack of oxygen and glycogen and the accumulation of lactic acid (a byproduct of ATP production in the absence of oxygen), together with the lack of ATP, all contribute to muscle fatigue.

Muscle metabolism

In order for muscles to contract, ATP must be available in the muscle fiber. ATP is available from the following sources:

■ Within the muscle fiber. ATP available within the muscle fiber can maintain muscle contraction for several seconds.

■ Creatine phosphate. Creatine phosphate, a high-energy molecule stored in muscle cells, transfers its high-energy phosphate group to ADP to form ATP. The creatine phosphate in muscle cells is able to generate enough ATP to maintain muscle contraction for about 15 seconds.

■ Glucose stored within the cell. Glucose within the cell is stored in the carbohydrate glycogen. Through the metabolic process of glycogenolysis, glycogen is broken down to release glucose. ATP is then generated from glucose by cellular respiration.

■ Glucose and fatty acids obtained from the bloodstream. When energy requirements are high, glucose from glycogen stored in the liver and fatty acids from fat stored in adipose cells and the liver are released into the bloodstream. Glucose and fatty acids are then absorbed from the bloodstream by muscle cells. ATP is then generated from these energy-rich molecules by cellular respiration.

Cellular respiration is the process by which ATP is obtained from energy-rich molecules. Several major metabolic pathways are involved, some of which require the presence of oxygen. Here's a summary of the important pathways:

■ In glycolysis, glucose is broken down to pyruvic acid, and two ATP molecules are generated even though oxygen is not present. The production of ATP without the use of oxygen is called anaerobic respiration, and, because no oxygen is used during the various metabolic steps of this pathway, glycolysis is called an anaerobic process.

■ During anaerobic respiration, pyruvic acid is converted to lactic acid. Lactic acid (via liver enzymes) can be converted back to pyruvic acid and, with the presence of oxygen, pyruvic acid can enter the mitochondria.

Anaerobic respiration has advantages and disadvantages:

- Advantages: Anaerobic respiration is relatively rapid, and it does not require oxygen.

- Disadvantages: Anaerobic respiration generates only two ATPs and produces lactic acid. Most lactic acid diffuses out of the cell and into the bloodstream and is subsequently absorbed by the liver. Some of the lactic acid remains in the muscle fibers, where it contributes to muscle fatigue. During strenuous exercise, a lot of ATP needs to be produced. Since a person is exercising faster than they are bringing in oxygen, the body tries to make ATP using the anaerobic pathway. This results in the production of ATP and lots of lactic acid. After exercise, the liver and muscles need to convert the lactic acid back to pyruvic acid. In order to do that, a lot of the oxygen the body is now taking in does the conversion instead of being used elsewhere. This is known as "repaying the debt," hence the term "oxygen debt."

- In aerobic respiration, pyruvic acid (from glycolysis) and fatty acids (from the bloodstream) are broken down, producing H_2O and CO_2 (carbon dioxide) and regenerating the coenzymes for glycolysis. A total of 36 ATP molecules are produced (including the two from glycolysis). However, oxygen is required for this pathway.

Aerobic respiration also has advantages and disadvantages:

- Advantages: Aerobic respiration generates a large amount of ATP.

- Disadvantages: Aerobic respiration is relatively slow and requires oxygen.

When the ATP generated from creatine phosphate is depleted, the immediate requirements of contracting muscle fibers force anaerobic respiration to begin. Anaerobic respiration can supply ATP for about 30 seconds. If muscle contraction continues, aerobic respiration, the slower ATP-producing pathway, begins and produces large amounts of ATP as long as oxygen is available. Eventually, oxygen is depleted, and aerobic respiration stops. However, ATP production by anaerobic respiration may still support some further muscle contraction. Ultimately, the accumulation of lactic acid from anaerobic respiration and the depletion of resources (ATP, oxygen, and glycogen) lead to muscle fatigue, and muscle contraction stops.

Structure of Cardiac and Smooth Muscle

Although it is striated, cardiac muscle differs from skeletal muscle in that it is highly branched with cells connected by overlapping projections of the sarcolemma called intercalated discs. These discs contain desmosomes and gap junctions. In addition, cardiac muscle is autorhythmic, generating its own action potential, which spreads rapidly throughout muscle tissue by electrical synapses across the gap junctions.

Due to its irregular arrangement of actin and myosin filaments, smooth muscle does not have the striated appearance of skeletal muscle. In addition, the sarcolemma does not form a system of transverse tubules. As a result, contraction is controlled and relatively slow—properties appropriate for smooth muscle function.

In addition to the thick myosin and thin actin filaments, smooth muscles possess noncontracting intermediate filaments. The intermediate fibers attach to dense bodies that are scattered through the sarcoplasm and attached to the sarcolemma. During contraction, the movements of myosin and actin are transferred to intermediate fibers, which pull on the dense bodies; these in turn pull the muscle cells together. In this way, the dense bodies function similarly to the Z discs in striated muscles.

Chapter Check-Out

Q&A

1. In anaerobic respiration, _____ _____ accumulates and leads to muscle fatigue.

2. True or False: Fast fibers contract rapidly, fatigue rapidly, and are highly vascularized.

3. Which of the following is true of glycolysis?
   a. Pyruvic acid is broken down to produce ATP.
   b. Cellular respiration is primarily utilized in a 100-meter dash.
   c. It is the slowest form of cellular respiration.
   d. It produces oxygen debt.
   e. Oxygen is vital for this type of cellular respiration.

4. True or False: In muscle contractions, the length of the muscle is always shortened.

5. The _____ _____ is the time immediately following a stimulus where the muscle will not respond to a second stimulus.

Answers: 1. lactic acid, **2.** F, **3.** b, **4.** F (isometric contractions), **5.** refractory period

Chapter 9

THE MUSCULAR SYSTEM

Chapter Check-In

❑ Characterizing how muscles are named

❑ Identifying the muscles associated with each region of the body

❑ Listing the various muscles of the body

Knowing the muscular organization of each region of the body is crucial in anatomy. With an understanding of where a muscle originates and inserts, you can calculate the movements that will occur at a joint when these two points are brought together following an isotonic muscular contraction. The orientation, placement, and coordination of these muscles allow the human body to produce a wide range of voluntary movements.

The muscular system consists of skeletal muscles and their associated connective tissues. It does not include cardiac muscle and smooth muscle, which are associated with the systems in which they are found, such as the cardiovascular, digestive, urinary, or other organ systems.

Skeletal Muscle Actions

A skeletal muscle is attached to one bone and extends across a joint to attach to another bone. A muscle can also attach a bone to another structure, such as skin. When the muscle contracts, one of the structures usually remains stationary, while the other moves. The following terms refer to this characteristic of muscle contraction:

- The *origin* of the muscle is the muscle end that attaches to the stationary structure, usually a bone or a bony structure.

- The *insertion* of the muscle is the muscle end that attaches to the moving structure.

- The *belly* of the muscle is that part of the muscle between the origin and insertion.

Several muscles usually influence a particular body movement:

- The prime mover is the muscle that is most responsible for the movement.

- **Synergists** are other muscles that assist the prime mover. Synergists may stabilize nearby bones or refine the movement of the prime mover.

- **Antagonists** are muscles that cause a movement opposite to that of the prime mover. For example, if the prime mover raises an arm, its antagonist pulls the arm down. An antagonist is generally attached to the opposite side of the joint to which the prime mover is attached.

Names of Skeletal Muscles

Skeletal muscles are often named after the following characteristics:

- *Number of origins:* Biceps, triceps, and quadriceps indicate two, three, and four origins, respectively.

- *Location of origin or insertion:* The sternocleidomastoid names the sternum ("sterno") and clavicle ("cleido") as its origins and the mastoid process of the temporal bone as its insertion.

- *Location:* In addition to its origin or insertion, a muscle name may indicate a nearby bone or body region. For example, the temporalis muscle covers the temporal bone.

- *Shape:* The deltoid (triangular), trapezius (trapezoid), serratus (sawtoothed), and rhomboideus major (rhomboid) muscles have names that describe their shapes.

- *Direction of muscle fibers:* The terms rectus (parallel), transverse (perpendicular), and oblique (at an angle) in muscle names refer to the direction of the muscle fibers with respect to the midline of the body.

- *Size:* Maximus (largest), minimus (smallest), longus (longest), and brevis (shortest) are common suffixes added to muscle names.

- *Action:* Terms such as flexor (flex the arm), extensor (extend the arm), abductor (move the arm laterally away from the torso), and adductor (return the arm to the torso) are added as prefixes to muscle names to indicate the kind of movement generated by the muscle.

Muscle Size and Arrangement of Muscle Fascicles

The size of a muscle influences its capabilities. When a muscle fiber (muscle cell) contracts, it can shorten to nearly half its relaxed length. The longer a muscle fiber, the greater range of movement it can generate. In contrast, an increase in the number of muscle fibers increases the strength of the contraction.

Muscle fibers are grouped into fascicles, which in turn are grouped to form a muscle. The size (length) and number of fascicles determine the strength and range of movement of a muscle. Common fascicle patterns include the following:

- *Parallel fascicles* have their long axes parallel to each other. Parallel fascicles can be flat or straplike, or they can bulge at their bellies and be spindle-shaped or fusiform.

- *Circular fascicles* are arranged in concentric rings. Muscles with this pattern form sphincter muscles that control the opening and closing of orifices.

- *Pennate fascicles* are short and attach obliquely to a long tendon that extends across the entire muscle. In a unipennate pattern, the muscle resembles one half of a feather (the tendon is represented by the shaft of the feather). A bipennate pattern resembles a complete feather, with fascicles attached to both sides of a central tendon. A multipennate pattern of fascicles resembles three or more feathers attached at their bases.

Major Skeletal Muscles

The major skeletal muscles are illustrated in Figures 9-1 through 9-6 and described in Tables 9-1 through 9-4.

Figure 9-1 The major skeletal muscles—anterior superficial view.

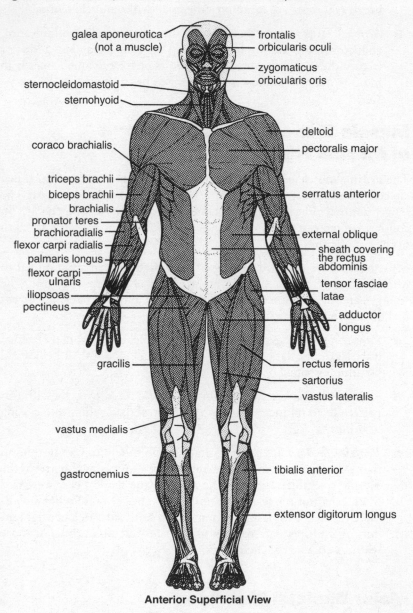

Anterior Superficial View

Figure 9–2 The major skeletal muscles—posterior superficial view.

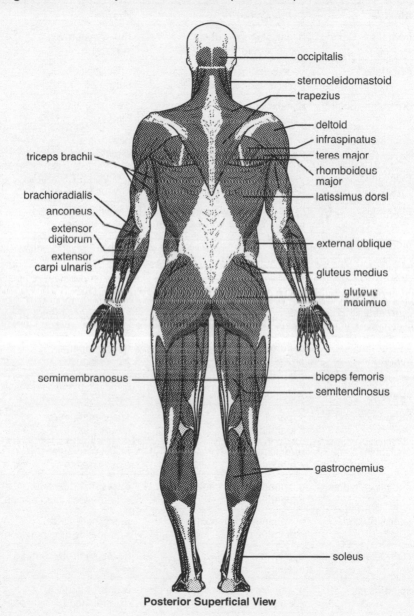

Posterior Superficial View

Table 9-1   Muscles of the Head and Neck

Muscle	Origin/Insertion	Action
frontalis	o: galea aponeurotica i: skin around eyes	raises eyebrows; surprised
occiptalis	o: occipital bone i: galea aponeurotica	pulls scalp back; surprised
orbicularis oculi	o: maxillary and frontal bones i: eyelids	closes eyelids; blinking
orbicularis oris	o: muscle fibers around mouth i: skin around mouth	closes lips; kissing
buccinator	o: maxilla and mandible i: orbicularis oris	compresses cheek; whistling
platysma	o: fascia in upper chest i: mandible and corner of mouth	lowers mandible; opens mouth
mentalis	o: mandible i: skin of chin	protrudes lower lip; pouting
risorius	o: fascia on masseter muscle i: skin at corner of mouth	lateral movement of lips; grimacing
zygomaticus	o: zygomatic bone i: skin around mouth	raises edges of mouth; smiling
levator labii superioris	o: infroarbital margin of maxilla i: skin of upper lip	raises upper lip; as in disgust
depressor labii inferioris	o: mandible i: skin of lower lip	lowers lower lip
temporalis	o: parietal bone i: mandible	raises mandible; closes mouth
masseter	o: zygomatic arch i: mandible	raises mandible; closes mouth
medial pterygoid	o: sphenoid and maxilla i: mandible	raises mandible; side-to-side mouth motion
lateral pterygoid	o: sphenoid and maxilla i: mandible	raises mandible; side-to-side mouth motion
sternocleidomastoid	o: sternum and clavicle i: temporal bone	flexes and rotates head
splenius capitis	o: cervical and thoracic vertebrae i: temporal bone	rotates, bends, or extends head

Muscle	Origin/Insertion	Action
semispinalis capitis	o: cervical and thoracic vertebrae i: occipital bone	rotates or extends head
longissimus	o: cervical and thoracic vertebrae i: temporal bone	rotates, bends, or extends head
omohyoid	o: scapula i: hyoid bone	depresses hyoid bone
sternohyoid	o: sternum and clavicle i: hyoid bone	depresses hyoid bone

Table 9-2 Muscles of the Neck, Shoulder, Thorax, and Abdominal Wall

Muscle	Origin/Insertion	Action
semispinalis capitis	o: cervical and thoracic vertebrae i: occipital bone	extends and rotates head
splenius capitis	o: cervical and thoracic vertebrae i: occipital and temporal bone	extends and rotates head
deltoid	o: clavicle and scapula i: humerus	abducts, flexes, extends, and rotates arm
pectoralis major	o: clavicle, sternum, ribs i: humerus	flexes, adducts, and rotates arm
infraspinatus	o: scapula i: humerus	rotates arm
teres major	o: scapula i: humerus	extends and rotates arm
latissimus dorsi	o: vertebrae, ribs, ilium i: humerus	extends, adducts, and rotates arm
levator scapulae	o: cervical vertebrae i: scapula	elevates scapula
pectoralis minor	o: ribs i: scapula	stabilizes scapula; elevates ribs
serratus anterior	o: ribs i: scapula	stabilizes scapula; elevates ribs

(continued)

Table 9-2 (continued)

Muscle	Origin/Insertion	Action
trapezius	o: occipital bone and cervical/ thoracic vertebrae i: scapula and clavicle	elevates, adducts, and rotates scapula
rhomboideous major/ rhomboideus minor	o: cervical/thoracic vertebrae i: scapula	adducts and rotates scapula
rectus abdominis	o: pubic crest and symphysis i: xiphoid process and ribs	flexes vertebral column; compresses abdomen
external oblique	o: ribs i: linea alba, ilium	compresses abdomen; rotates trunk
transverse abdominis	o: ilium, ribs i: linea alba, xiphoid process	compresses abdomen
external intercostals	o: lower border of rib above i: upper border of rib below	elevates ribs; aids inspiration
internal intercostals	o: upper border of rib below i: lower border of rib above	pulls ribs together; aids expiration
diaphragm	o: lower ribs, sternum i: central tendon	aids inspiration
spinalis	o: lumbar and thoracic vertebrae i: thoracic and cervical vertebrae	extends vertebral column
longissimus	o: lumbar and cervical vertebrae i: temporal bone, vertebrae	extends vertebral column
iliocostalis	o: ilium, ribs i: ribs	extends vertebral column

Figure 9-3 The major skeletal muscles—anterior and lateral views.

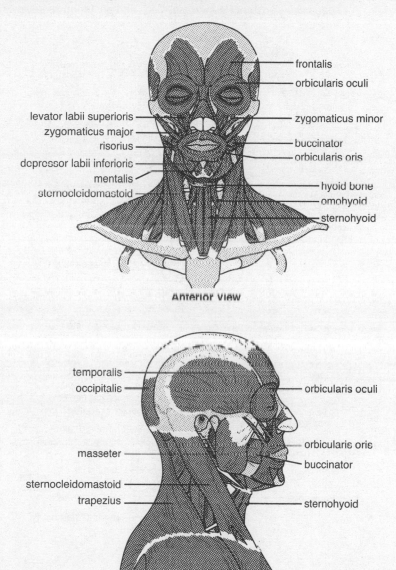

Figure 9-4 The major skeletal muscles—anterior superficial view, anterior deep view, posterior superficial view, and posterior deep view.

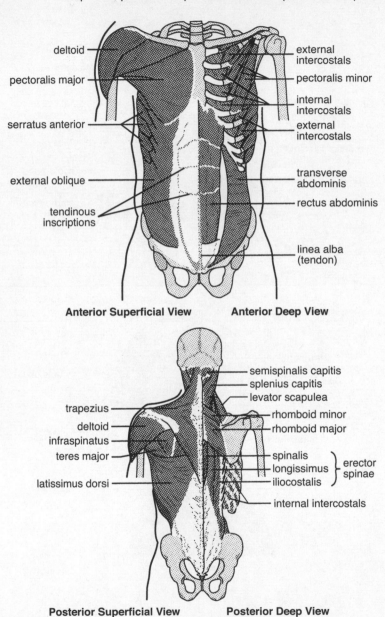

Anterior Superficial View   Anterior Deep View

Posterior Superficial View   Posterior Deep View

Table 9-3 Muscles of the Arm and Forearm

Muscle	Origin/Insertion	Action
coracobrachiolis	o: scapula i: humerus	flexes and adducts arm
biceps brachii	o: scapula, glenoid cavity i: radius	flexes arm, flexes forearm, rotates hand
brachialis	o: humerus i: ulna	flexes forearm
brachioradialis	o: humerus I: radlus	flexes forearm
triceps brachii	o: humerus i: ulna	extends forearm
anconeus	o: humerus i: ulna	extends forearm
pronator teres	o: humerus, ulna i: radius	medial rotation of forearm pronate hand
pronator quadratus	o: ulna I: radius	medial rotation of forearm pronate hand
supinator	o: ulna i: radius	lateral rotation of forearm
flexor carpi radialis	o: humerus i: metacarpals	flexes and abducts wrist
flexor carpi ulnaris	o: humerus, ulna i: carpals, metacarpals	flexes and abducts wrist
flexor digitorum superficialis	o: humerus, ulna, radius i: phalanges	flexes fingers 2–5
flexor digitorum profundus	o: ulna i: phalanges	flexes distal fingers 2–5
palmaris longus	o: humerus i: flexor retinaculum	flexes wrist

(continued)

Table 9-3 (continued)

Muscle	Origin/Insertion	Action
extensor carpi radialis longus	o: humerus i: second metacarpal	extends and abducts wrist
extensor carpi ulnaris	o: humerus, ulna i: fifth metacarpal	extends and adducts wrist
extensor digitorum	o: humerus i: distal phalanges	extends fingers 2–5
extensor pollicis brevis	o: radius i: phalanx of thumb	extends thumb
extensor pollicis longus	o: radius i: phalanx of thumb	extends thumb
extensor indicis	o: ulna i: index finger	extends index finger
abductor pollicis longus	o: radius and thumb i: first metacarpal	abducts and extends thumb

Table 9-4 Muscles of the Thigh and Leg

Muscle	Origin/Insertion	Action
gluteus maximus	o: ilium, sacrum, coccyx i: femur	extends and rotates thigh
gluteus medius	o: ilium i: femur	abducts and rotates thigh
pectineus	o: pubis i: femur	adducts and flexes thigh
adductor longus	o: pubis i: femur	adducts, flexes, and rotates thigh
adductor brevis	o: pubis i: femur	adducts, flexes, and rotates thigh
adductor magnus	o: pubis, ischium i: femur	adducts, flexes, and rotates thigh

Muscle	Origin/Insertion	Action
gracilis	o: pubis i: tibia	adducts thigh and flexes leg
sartorius	o: ilium i: tibia	flexes and rotates thigh; flexes leg
quadriceps femoris: rectus femoris	o: ilium, femur i: patella, tibia	extends leg; flexes thigh
quadriceps femoris: vastus lateralis	o: ilium, femur i: patella, tibia	extends leg
quadriceps femoris: vastus medialis	o: ilium, femur i: patella, tibia	extends leg
quadriceps femoris: vastus intermedius	o: Illum, femur i: patella, tibia	extends leg
biceps femoris (hamstrings)	o: ischium, femur i: fibula	flexes and rotates leg; extends thigh
semitendinosus (hamstrings)	o: ischium i: tibia	flexes and rotates leg; extends thigh
semimembranosus (hamstrings)	o: ischium i: tibia	flexes and rotates leg; extends thigh
tibialis anterior	o: tibia i: 1st metatarsal and cunciform	dorsiflexes and inverts foot
extensor digitorum longus	o: tibia, fibula i: phalanges of toes	dorsiflexes and everts foot; extends toes
gastrocnemius	o: femur i: calcanues	plantar flexes foot; flexes leg
soleus	o: tibia, fibula i: calcaneus	plantar flexes foot

Figure 9-5 The major skeletal muscles—anterior superficial view, anterior deep view, posterior superficial view, and posterior deep view.

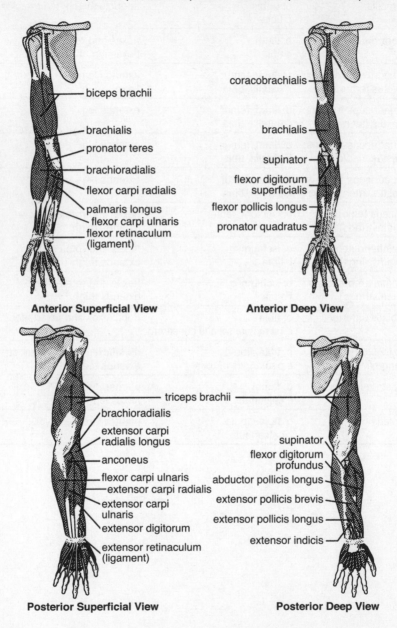

biceps brachii

brachialis

pronator teres

brachioradialis

flexor carpi radialis

palmaris longus

flexor carpi ulnaris

flexor retinaculum (ligament)

Anterior Superficial View

coracobrachialis

brachialis

supinator

flexor digitorum superficialis

flexor pollicis longus

pronator quadratus

Anterior Deep View

triceps brachii

brachioradialis

extensor carpi radialis longus

anconeus

flexor carpi ulnaris

extensor carpi radialis

extensor carpi ulnaris

extensor digitorum

extensor retinaculum (ligament)

Posterior Superficial View

supinator

flexor digitorum profundus

abductor pollicis longus

extensor pollicis brevis

extensor pollicis longus

extensor indicis

Posterior Deep View

Figure 9-6   The major skeletal muscles—anterior superficial view, anterior deep view, posterior superficial view, and posterior deep view.

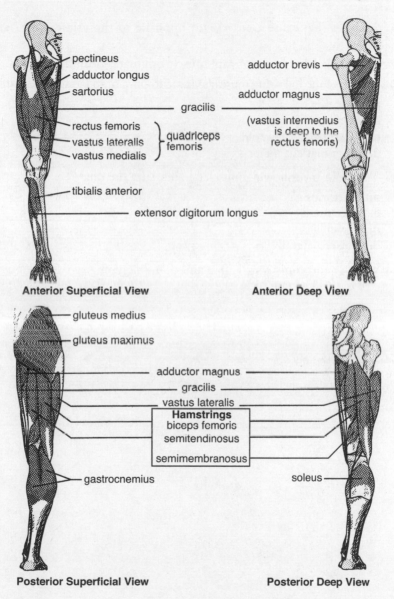

pectineus

adductor longus

sartorius

gracilis

rectus femoris

vastus lateralis } quadriceps femoris

vastus medialis

tibialis anterior

extensor digitorum longus

adductor brevis

adductor magnus

(vastus intermedius is deep to the rectus femoris)

Anterior Superficial View

Anterior Deep View

gluteus medius

gluteus maximus

adductor magnus

gracilis

vastus lateralis

Hamstrings
biceps femoris
semitendinosus
semimembranosus

gastrocnemius

soleus

Posterior Superficial View

Posterior Deep View

Chapter Check-Out

Q&A

1. Muscles that cause a movement opposite to the prime mover are called _____.

2. True or False: The omohyoid muscle originates from the scapula.

3. Which of the following muscle(s) is/are important in facial expression?
   a. risorius
   b. teres major
   c. sternocleidomastoid
   d. zygomaticus major

4. Which of the following muscle(s) adduct(s) the thigh?
   a. sartorius
   b. pectineus
   c. gracilis
   d. vastus lateralis

Answers: 1. antagonists, **2.** T, **3.** a and d, **4.** b and c

Chapter 10

NERVOUS TISSUE

Chapter Check-In

❑ Naming the basic parts of a neuron

❑ Classifying neurons according to function and structure

❑ Listing the different types of neuroglia that support and protect neurons

❑ Describing how a neural impulse is generated and moves across a synapse

The human nervous system accomplishes countless actions that are crucial to existence. The functional cellular unit of this complex system is the neuron. Moreover, the primary goal of neuroscience is to understand the complexities of the nervous system in terms of the morphology and physiology of the neuron and its support cells, the neuroglia.

Nervous tissue consists of two kinds of nerve cells—neurons and neuroglia.

Neurons

A neuron is a cell that transmits nerve impulses. It consists of the following parts, shown in Figure 10-1:

- The cell body (soma or perikaryon) contains the nucleus and other cell organelles.

- There are clusters of rough endoplasmic reticulum (not shown in Figure 10-1) that are called Nissl bodies or are sometimes referred to as chromatophilic substances.

■ The dendrite is typically a short, abundantly branched, slender process (extension) of the cell body that receives stimuli.

■ The **axon** is typically a long, slender process of the cell body that sends nerve impulses. It emerges from the cell body at the cone-shaped axon hillock. Nerve impulses arise in the trigger zone, generally located in the initial segment, an area just outside the axon hillock. The cytoplasm of the axon, the **axoplasm,** is surrounded by its plasma membrane, the **axolemma.** A few axons branch along their lengths to form axon collaterals, and these branches may return to merge with the main axon. At its end, each axon or axon collateral usually forms numerous branches (**telodendria**), with most branches terminating in bulb-shaped structures called synaptic knobs (synaptic end bulbs, also called terminal boutons). The synaptic knobs contain neurotransmitters, chemicals that transmit nerve impulses to a muscle or another neuron.

Figure 10-1 Parts of a neuron.

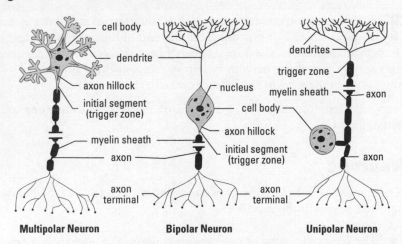

Multipolar Neuron   **Bipolar Neuron**   **Unipolar Neuron**

Neurons can be classified by function or by structure. Functionally, they fall into three groups:

■ *Sensory neurons* (**afferent** neurons) transmit sensory impulses from the skin and other sensory organs or from various places within the body toward the central nervous system (CNS), which consists of the brain and spinal cord.

■ *Motor neurons* (**efferent** neurons) transmit nerve impulses from the CNS toward effectors, target cells that produce some kind of response. Effectors include muscles, sweat glands, and many other organs.

■ *Association neurons* (**interneurons**) are located in the CNS and transmit impulses from sensory neurons to motor neurons. More than 90 percent of the neurons of the body are association neurons.

Neurons are structurally classified into three groups, as shown in Figure 10-1:

■ *Multipolar neurons* have one axon and several to numerous dendrites. Most neurons are of this type.

■ *Bipolar neurons* have one axon and one dendrite. They emerge from opposite sides of the cell body. Bipolar neurons are found only as specialized sensory neurons in the eye, ear, or olfactory organs.

■ *Unipolar neurons* have one process of emerging from the cell body that branches, T-fashion, into two processes. Both processes function together as a single axon. Dendrites emerge from one of the terminal ends of the axon. The trigger zone in a unipolar neuron is located at the junction of the axon and dendrites. Unipolar neurons are mostly sensory neurons.

The following terms apply to neurons and groups of neurons:

■ A nerve fiber is an axon.

■ A nerve is a bundle of nerve fibers in the peripheral nervous system (PNS). Most nerves contain both sensory and motor fibers. Cell bodies are usually grouped into separate bundles called ganglia.

■ A peripheral nerve consists of three layers:

■ The *epineurium* is the outer layer that surrounds the entire nerve.

■ The *perineurium* surrounds bundles of axons. Bundles of axons are called fascicles. There could be 10 or more fascicles per nerve.

■ Surrounding each individual axon is the *endoneurium*.

■ A nerve tract is a bundle of nerve fibers in the CNS.

Neuroglia

Neuroglia (glia) are cells that support and protect neurons. The following four neuroglia are found in the CNS:

■ **Astrocytes** have numerous processes that give the cell a star-shaped appearance. Astrocytes maintain the ion balance around neurons and control the exchange of materials between blood vessels and neurons.

■ *Oligodendrocytes* have fewer processes than astrocytes. They wrap these cytoplasmic processes around neurons to create an insulating barrier called a myelin sheath.

■ *Microglia* are phagocytic macrophages that provide a protective function by engulfing microorganisms and cellular debris.

■ *Ependymal cells* line the fluid-filled cavities of the brain and spinal cord. Many are ciliated.

Two kinds of neuroglia are found in PNS:

■ *Schwann cells* (neurolemmocytes) wrap around axons to produce an insulating myelin sheath. Schwann cells provide the same function in the PNS as oligodendrocytes provide in the CNS.

■ *Satellite cells* are located in ganglia, where they surround the cell bodies of neurons.

Myelination

In the PNS, the myelin sheath is an insulation formed by Schwann cells around axons (refer to Figure 10-1). Each Schwann cell tightly wraps around the axon numerous times to form a multilayered insulation. The last wrapping of the plasma membrane of the Schwann cell, the neurilemma, is loose and contains the nucleus, cytoplasm, and organelles of the Schwann cell.

The myelin sheath consists of numerous Schwann cell wrappings along the length of the axon. Spaces occur between adjacent Schwann cells, leaving uninsulated areas, or neurofibral nodes (nodes of Ranvier), along the axons. As an insulator, the Schwann cells interrupt the continuous conduction of a nerve impulse along the axon. A signal is transmitted by local current in the interior of the axon. Axon potentials are generated only at the nodes. This gives the appearance of "jumping" (salutatory conduction) from node to node. In this fashion, the myelin sheath provides insulation between adjacent nerve fibers, preventing the cross-over of one nerve impulse to an adjacent axon.

In the CNS, the axons of neurons are insulated by oligodendrocytes. The myelin sheath is formed when a process extended by the oligodendrocyte wraps around an axon. Many axons may be myelinated by multiple processes from a single oligodendrocyte.

The white and gray matter of the brain and spinal cord are distinguished by the presence or absence of myelin sheaths:

- White matter contains the myelinated axons of neurons. (The white color is from the myelin sheaths.)

- Gray matter contains the unmyelinated portions of neurons (cell bodies, dendrites, and axon terminals), unmyelinated axons, and neuroglia.

Transmission of Nerve Impulses

The transmission of a nerve impulse along a neuron from one end to the other occurs as a result of electrical changes across the membrane of the neuron. The membrane of an unstimulated neuron is polarized—that is, there is a difference in electrical charge between the outside and inside of the membrane. The inside is negative with respect to the outside.

Polarization is established by maintaining an excess of sodium ions (Na^+) on the outside and an excess of potassium ions (K^+) on the inside. A certain amount of Na^+ and K^+ is always leaking across the membrane through leakage channels, but Na^+/K^+ pumps in the membrane actively restore the ions to the appropriate side.

The main contribution to the resting membrane potential (a polarized nerve) is the difference in permeability of the resting membrane to potassium ions versus sodium ions. The resting membrane is much more permeable to potassium ions than to sodium ions resulting in slightly more net potassium ion diffusion (from the inside of the neuron to the outside) than sodium ion diffusion (from the outside of the neuron to the inside) causing the slight difference in polarity right along the membrane of the axon.

Other ions, such as large, negatively charged proteins and nucleic acids, reside within the cell. It is these large, negatively charged ions that contribute to the overall negative charge on the inside of the cell membrane as compared to the outside.

In addition to crossing the membrane through leakage channels, ions may cross through **gated channels.** Gated channels open in response to neurotransmitters, changes in membrane potential, or other stimuli.

The following events characterize the transmission of a nerve impulse (see Figure 10-2):

■ **Resting potential.** The resting potential describes the unstimulated, polarized state of a neuron (at about −70 millivolts).

■ **Graded potential.** A graded potential is a change in the resting potential of the plasma membrane in the response to a stimulus. A graded potential occurs when the stimulus causes Na⁺ or K⁺ gated channels to open. If Na⁺ channels open, positive sodium ions enter, and the membrane depolarizes (becomes more positive). If the stimulus opens K⁺ channels, then positive potassium ions exit across the membrane and the membrane **hyperpolarizes** (becomes more negative). A graded potential is a local event that does not travel far from its origin. Graded potentials occur in cell bodies and dendrites. Light, heat, mechanical pressure, and chemicals, such as neurotransmitters, are examples of stimuli that may generate a graded potential (depending upon the neuron).

Figure 10-2 Events that characterize the transmission of a nerve impulse.

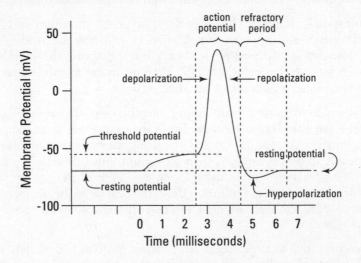

The following four steps describe the initiation of an impulse to the "resetting" of a neuron to prepare for a second stimulation:

1. **Action potential.** Unlike a graded potential, an action potential is capable of traveling long distances. If a depolarizing graded potential is sufficiently large, Na^+ channels in the trigger zone open. In response, Na^+ on the outside of the membrane becomes depolarized (as in a graded potential). If the stimulus is strong enough—that is, if it is above a certain threshold level—additional Na^+ gates open, increasing the flow of Na^+ even more, causing an action potential, or complete depolarization (from −70 to about +30 millivolts). This in turn stimulates neighboring Na^+ gates, farther down the axon, to open. In this manner, the action potential travels down the length of the axon as opened Na^+ gates stimulate neighboring Na^+ gates to open. The action potential is an all-or-nothing event: When the stimulus fails to produce depolarization that exceeds the threshold value, no action potential results, but when threshold potential is exceeded, complete depolarization occurs.

2. Repolarization. In response to the inflow of Na^+, K^+ channels open, this time allowing K^+ on the inside to rush out of the cell. The movement of K^+ out of the cell causes repolarization by restoring the original membrane polarization. Unlike the resting potential, however, in repolarization the K^+ are on the outside and the Na^+ are on the inside. Soon after the K^+ gates open, the Na^+ gates close.

3. **Hyperpolarization.** By the time the K^+ channels close, more K^+ have moved out of the cell than is actually necessary to establish the original polarized potential. Thus, the membrane becomes hyperpolarized (about −80 millivolts).

4. Refractory period. With the passage of the action potential, the cell membrane is in an unusual state of affairs. The membrane is polarized, but the Na^+ and K^+ are on the wrong sides of the membrane. During this refractory period, the axon will not respond to a new stimulus. To reestablish the original distribution of these ions, the Na^+ and K^+ are returned to their resting potential location by Na^+/K^+ pumps in the cell membrane. Once these ions are completely returned to their resting potential location, the neuron is ready for another stimulus.

The Synapse

A **synapse,** or synaptic cleft, is the gap that separates adjacent neurons or a neuron and a muscle. Transmission of an impulse across a synapse, from presynaptic cell to postsynaptic cell, is chemical. In chemical synapses, action potentials are transferred across the synapse by the diffusion of chemicals, as follows:

1. Calcium (Ca^{2+}) gates open. When an action potential reaches the end of an axon, the depolarization of the membrane causes gated channels to open that allow Ca^{2+} to enter.

2. Synaptic vesicles release a neurotransmitter. The influx of Ca^{2+} into the terminal end of the axon causes synaptic vesicles to merge with the presynaptic membrane, releasing a neurotransmitter into the synaptic cleft.

3. The neurotransmitter binds with postsynaptic receptors. The neurotransmitter diffuses across the synaptic cleft and binds with specialized protein receptors on the postsynaptic membrane. Different proteins are receptors for different neurotransmitters.

4. The postsynaptic membrane is excited or inhibited. Depending upon the kind of neurotransmitter and the kind of membrane receptor, there are two possible outcomes for the postsynaptic membrane, both of which are graded potentials:

   ■ If positive ion gates open (allowing more Na^+ and Ca^{2+} to enter than K^+ to exit), the membrane becomes depolarized, which results in an excitatory postsynaptic potential (EPSP). If the threshold potential is exceeded, an action potential is generated.

   ■ If K^+ or chlorine ion (Cl^-) gates open (allowing K^+ to exit or Cl^- to enter), the membrane becomes more polarized (hyperpolarized), which results in an inhibitory postsynaptic potential (IPSP). As a result, it becomes more difficult to generate an action potential on this membrane.

5. The neurotransmitter is degraded and recycled. After the neurotransmitter binds to the postsynaptic membrane receptors, it is either transported back to and reabsorbed by the secreting neuron or broken down by enzymes in the synaptic cleft that come from the postsynaptic membrane. For example, the common neurotransmitter acetylcholine (ACh) is broken down by acetylcholinesterase (AChE). Reabsorbed and degraded neurotransmitters are recycled by the structures in the presynaptic area.

Here are some of the common neurotransmitters and the kinds of activity they generate:

- **Acetylcholine (ACh)** is commonly secreted at neuromuscular junctions, the gaps between motor neurons and skeletal muscle cells, where it stimulates muscles to contract by opening gated positive ion channels.

- Epinephrine, norepinephrine (NE), dopamine, and serotonin are derived from amino acids and are secreted mostly between neurons of the CNS. Norepinephrine is also found in the peripheral nervous system (PNS).

- Gamma aminobutyric acid (GABA) is usually an inhibitory neurotransmitter (opening gated Cl⁻ channels) among neurons in the brain.

Chapter Check-Out

Q&A

1. _____ neurons transmit nerve impulses away from the central nervous system (CNS).

2. True or False: Gray matter contains the myelinated axons of neurons.

3. True or False: The electrical charge of an unstimulated neuron is negative inside the membrane.

4. Clusters of rough endoplasmic reticulum, or sites of protein synthesis within a neuron, are called _____ _____.

Answers: 1. Efferent (or Motor), **2.** F, **3.** T, **4.** Nissl bodies

Chapter 11

THE NERVOUS SYSTEM

Chapter Check-In

❑ Classifying the components of the peripheral and central nervous systems

❑ Discovering the different regions of the brain and their functions

❑ Describing the circulation and production of cerebrospinal fluid

❑ Appreciating the organization and distribution of the spinal nerves

❑ Distinguishing between the sympathetic and parasympathetic components of the autonomic nervous system

The nervous system integrates and monitors the countless actions occurring simultaneously throughout the entire human body; therefore, every task a person accomplishes, no matter how menial, is a direct result of the components of the nervous system. These actions can be under voluntary control, like touching a computer key, or can occur without your direct knowledge, like digesting food, releasing enzymes from the pancreas, or other unconscious acts.

It is difficult to understand all the complexities of the nervous system because the field of neuroscience has rapidly evolved over the past 20 years; moreover, answers to new questions are being found almost daily. A thorough knowledge of the individual components of the nervous system and their functions, however, will lead you to a better understanding of how the human body works and will facilitate your future acquisition of knowledge about the nervous system.

Nervous System Organization

The nervous system consists of two parts, shown in Figure 11-1:

- The central nervous system (CNS) consists of the brain and spinal cord.

- The peripheral nervous system (PNS) consists of nerves outside the CNS.

Nerves of the PNS are classified in three ways. First, PNS nerves are classified by how they are connected to the CNS. Cranial nerves originate from or terminate in the brain, while spinal nerves originate from or terminate at the spinal cord.

Second, nerves of the PNS are classified by the direction of nerve propagation. Sensory (**afferent**) neurons transmit impulses from skin and other sensory organs or from various places within the body to the CNS. Motor (**efferent**) neurons transmit impulses from the CNS to effectors (muscles or glands).

Third, motor neurons are further classified according to the effectors they target. The **somatic nervous system (SNS)** directs the contraction of skeletal muscles. The **autonomic nervous system (ANS)** controls the activities of organs, glands, and various involuntary muscles, such as cardiac and smooth muscles.

The autonomic nervous system has two divisions:

- The sympathetic nervous system is involved in the stimulation of activities that prepare the body for action, such as increasing the heart rate, increasing the release of sugar from the liver into the blood, and other activities generally considered as fight-or-flight responses (responses that serve to fight off or retreat from danger).

- The parasympathetic nervous system activates tranquil functions, such as stimulating the secretion of saliva or digestive enzymes into the stomach and small intestine.

Generally, both sympathetic and parasympathetic systems target the same organs, but often work antagonistically. For example, the sympathetic system accelerates the heartbeat, while the parasympathetic system slows the heartbeat. Each system is stimulated as is appropriate to maintain homeostasis.

Figure 11-1 Two parts of the nervous system.

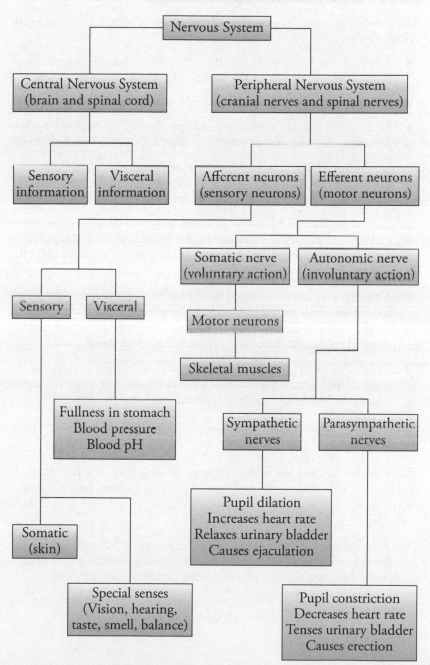

Nervous System Terminology

The following terms are commonly used in descriptions of nervous system features:

■ A nerve fiber is an axon or dendrite. A nerve is a bundle of nerve fibers in the PNS. A nerve tract is a bundle of nerve fibers in the CNS.

■ White matter consists of myelinated axons of neurons in the CNS.

■ Gray matter consists of unmyelinated portions of neurons (cell bodies, dendrites, and axon terminals), unmyelinated neurons, and neuroglia in the CNS.

■ Nuclei are clusters of cell bodies in the CNS. Ganglia are clusters of cell bodies in the PNS (except the **basal ganglia,** which are more appropriately called basal nuclei).

■ Vesicles are fluid-filled cavities in the brain that form during early development. The tissues that form the vesicles divide to become the various components of the brain.

■ Ventricles are interconnected cavities in the mature brain that originate from the fluid-filled vesicles. Circulating fluid (cerebrospinal fluid—CSF) in the ventricles provides nourishment for nervous tissue and transports waste away from the nervous tissue.

■ Peduncles are large tracts that emerge from certain regions of the brain. Their large size gives the appearance of supporting the structure from which they emerge (peduncle means "little foot").

The Brain

Three cavities, called the primary brain vesicles, form during the early embryonic development of the brain. These are the **forebrain** (prosencephalon), the **midbrain** (mesencephalon), and the **hindbrain** (rhombencephalon).

During subsequent development, the three primary brain vesicles develop into five secondary brain vesicles. The names of these vesicles and the major adult structures that develop from the vesicles follow (see Table 11-1):

■ The telencephalon generates the cerebrum (which contains the cerebral cortex, white matter, and basal ganglia).

- The diencephalon generates the thalamus, hypothalamus, and pineal gland.

- The mesencephalon generates the midbrain portion of the brainstem.

- The metencephalon generates the pons portion of the brainstem and the cerebellum.

- The myelencephalon generates the medulla oblongata portion of the brainstem.

Table 11-1   The Vesicles and Their Components

Primary Vesicles	Secondary Vesicles	Adult Structure	Important Components or Features
prosencephalon (forebrain)	telencephalon	cerebrum (cerebral hemispheres)	cerebral cortex (gray matter): motor areas, sensory areas, association areas
prosencephalon (forebrain)	telencephalon	cerebrum (cerebral hemispheres)	cerebral white matter: association fibers, commisural fibers, projection fibers
prosencephalon (forebrain)	telencephalon	cerebrum (cerebral hemispheres)	basal ganglia (gray matter): caudate nucleus and amygdala, putamen, globus pallidus
prosencephalon (forebrain)	diencephalon	diencephalon	thalamus: relays sensory information
prosencephalon (forebrain)	diencephalon	diencephalon	hypothalamus: maintains body homeostasis
prosencephalon (forebrain)	diencephalon	diencephalon	mammillary bodies: relays sensations of smells to cerebrum
prosencephalon (forebrain)	diencephalon	diencephalon	optic chiasma: crossover of optic nerves
prosencephalon (forebrain)	diencephalon	diencephalon	infundibulum: stalk of pituitary gland
prosencephalon (forebrain)	diencephalon	diencephalon	pituitary gland: source of hormones

(continued)

Table 11-1 *(continued)*

Primary Vesicles	Secondary Vesicles	Adult Structure	Important Components or Features
prosencephalon (forebrain)	diencephalon	diencephalon	epithalamus: pineal gland
mesencephalon (midbrain)	mesencephalon	brainstem	midbrain: cerebral peduncles, superior cerebellar peduncles, corpora quadrigemina, superior colliculi
rhombencephalon (hindbrain)	metencephalon	brainstem	pons: middle cerebellar peduncles, pneumotaxic area, apneustic area
rhombencephalon (hindbrain)	metencephalon	cerebellum	superior cerebellar peduncles, middle cerebellar peduncles, inferior cerebellar peduncles
rhombencephalon (hindbrain)	myelencephalon	brainstem	medulla oblongata: pyramids, cardiovascular center, respiratory center

- The cerebrum consists of two cerebral hemispheres connected by a bundle of nerve fibers, the **corpus callosum.** The largest and most visible part of the brain, the cerebrum, appears as folded ridges and grooves, called convolutions. The following terms are used to describe the convolutions:

  - A gyrus (plural, gyri) is an elevated ridge.

  - A sulcus (plural, sulci) is a shallow groove.

  - A fissure is a deep groove.

  The deeper fissures divide the cerebrum into five lobes (see Figure 11-2; most lobes are named after bordering skull bones): the frontal lobe, the parietal lobe, the temporal lobe, the occipital lobe, and the insula. All but the insula are visible from the outside surface of the brain.

Figure 11-2 Anatomy of the adult brain.

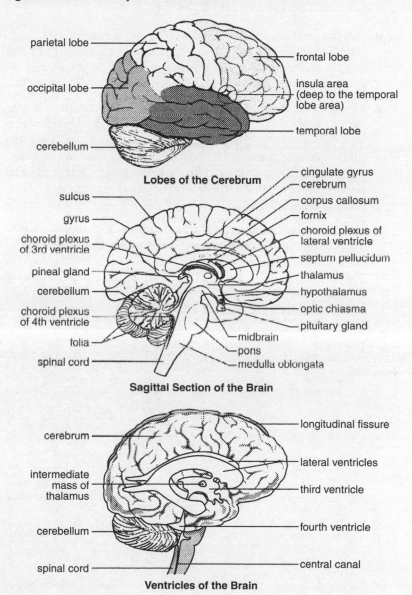

parietal lobe

frontal lobe

occipital lobe

insula area
(deep to the temporal
lobe area)

temporal lobe

cerebellum

Lobes of the Cerebrum

sulcus

gyrus

choroid plexus
of 3rd ventricle

pineal gland

cerebellum

choroid plexus
of 4th ventricle

folia

spinal cord

cingulate gyrus
cerebrum

corpus callosum

fornix

choroid plexus of
lateral ventricle

septum pellucidum

thalamus

hypothalamus

optic chiasma

pituitary gland

midbrain

pons

medulla oblongata

Sagittal Section of the Brain

cerebrum

intermediate
mass of
thalamus

cerebellum

spinal cord

longitudinal fissure

lateral ventricles

third ventricle

fourth ventricle

central canal

Ventricles of the Brain

A cross section of the cerebrum shows three distinct layers of nervous tissue (see the list below and Figure 11-3):

■ The cerebral cortex is a thin outer layer of gray matter. Such activities as speech, evaluation of stimuli, conscious thinking, and control of skeletal muscles occur here. These activities are grouped into motor areas, sensory areas, and association areas.

■ The cerebral white matter underlies the cerebral cortex. It contains mostly myelinated axons that connect cerebral hemispheres (association fibers), connect gyri within hemispheres (commissural fibers), or connect the cerebrum to the spinal cord (projection fibers). The corpus callosum is a major assemblage of association fibers that forms a nerve tract that connects the two cerebral hemispheres.

■ **Basal ganglia** (basal nuclei) are several pockets of gray matter located deep inside the cerebral white matter. The major regions in the basal ganglia—the caudate nuclei, the putamen, and the globus pallidus—are involved in relaying and modifying nerve impulses passing from the cerebral cortex to the spinal cord. Arm swinging while walking, for example, is controlled here.

Figure 11-3 The cerebral cortex and basal ganglia.

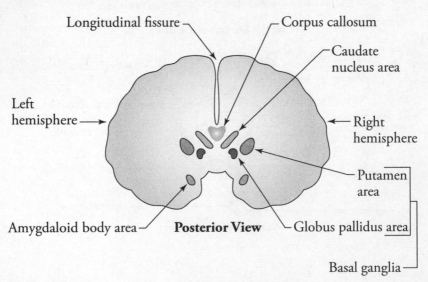

Longitudinal fissure

Corpus callosum

Caudate nucleus area

Left hemisphere

Right hemisphere

Putamen area

Amygdaloid body area

Posterior View

Globus pallidus area

Basal ganglia

■ The diencephalon connects the cerebrum to the brainstem. It consists of the following major regions:

■ The thalamus is a relay station for sensory nerve impulses traveling from the spinal cord to the cerebrum. Some nerve impulses are sorted and grouped here before being transmitted to the cerebrum. Certain sensations, such as pain, pressure, and sensitivity to temperature, are also evaluated here.

■ The epithalamus contains the pineal gland. The pineal gland secretes melatonin, a hormone that helps regulate the biological clock (sleep-wake cycles).

■ The hypothalamus regulates numerous important body activities. It controls the autonomic nervous system and regulates emotion, behavior, hunger, thirst, body temperature, and the biological clock. It also produces two hormones (antidiuretic hormone or ADH, and oxytocin) and various releasing hormones that control hormone production in the anterior pituitary gland.

The following structures are either included or associated with the hypothalamus:

■ The mammillary bodies relay information related to eating, such as chewing and swallowing

■ The infundibulum connects the pituitary gland to the hypothalamus.

■ The optic chiasma passes between the hypothalamus and the pituitary gland. Here, portions of the optic nerve from each eye cross over to the cerebral hemisphere on the opposite side.

■ The **brainstem** connects the diencephalon to the spinal cord. The brainstem resembles the spinal cord in that both consist of white matter fiber tracts surrounding a core of gray matter. The brainstem consists of the following four regions, all of which provide connections between various parts of the brain and between the brain and the spinal cord. (Some prominent structures of the brainstem regions are listed in Table 11-2 and illustrated in Figure 11-4, which also illustrates the relationship of the cranial nerves to the brainstem.)

Table 11-2   Structures of the Brainstem

Structure	Location
midbrain	uppermost part of the brainstem
pons	bulging region in the middle of the brainstem
medulla oblongata (medulla)	lower portion of the brainstem that merges with the spinal cord at the foramen magnum
reticular formation (small clusters of gray matter)	within the white matter of the various regions of the brainstem and certain regions of the spinal cord, diencephalon, and cerebellum

Figure 11-4 Prominent structures of the brainstem.

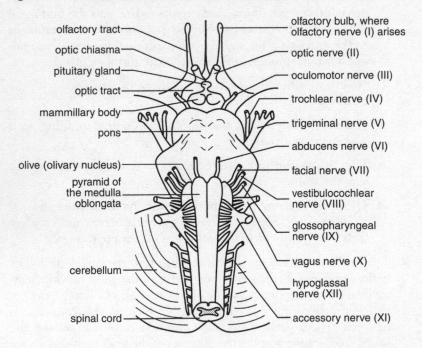

The **reticular activation system (RAS),** one component of the reticular formation, is responsible for maintaining wakefulness and alertness and for filtering out unimportant sensory information. Other components of the reticular formation are responsible for maintaining muscle tone and regulating visceral motor muscles.

- The cerebellum consists of a central region (the vermis) and two winglike lobes (the cerebellar hemispheres). Like that of the cerebrum, the surface of the cerebellum is convoluted, but the gyri, called folia, are parallel and give a pleated appearance. The cerebellum evaluates and coordinates motor movements by comparing actual skeletal movements to the movement that was intended.

The limbic system is a network of neurons that extends over a wide range of areas of the brain. The limbic system imposes an emotional aspect to behaviors, experiences, and memories. Emotions such as pleasure, fear, anger, sorrow, and affection are imparted to events and experiences. The limbic system accomplishes this by a system of fiber tracts (white matter) and gray matter that pervades the diencephalon and encircles the inside border of the cerebrum. The following components are also illustrated in Figure 11-5:

- The hippocampus (located in the temporal lobes of the cerebral hemispheres)

- The dentate gyrus (located in the cerebral hemisphere, within the hippocampus)

- The **amygdala** or amygdaloid body (an almond-shaped body associated with the caudate nucleus of the basal ganglia)

- The mammillary bodies

- The anterior thalamic nuclei (in the thalamus)

- The fornix (a bundle of fiber tracts that links components of the limbic system)

Figure 11-5 Features of the limbic system.

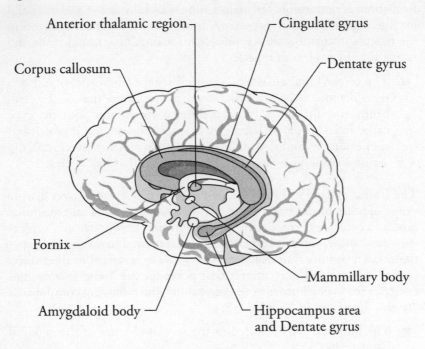

The Ventricles and Cerebrospinal Fluid

There are four cavities in the brain, called ventricles. The ventricles are filled with cerebrospinal fluid (CSF), which provides the following functions:

- Absorbs physical shocks to the brain

- Distributes nutritive materials to and removes wastes from nervous tissue

- Provides a chemically stable environment

There are four ventricles:

- Each of two lateral ventricles (ventricles 1 and 2) occupies a cerebral hemisphere.

- The third ventricle is connected by a passage (interventricular foramen) to each of the two lateral ventricles.

■ The fourth ventricle connects to the third ventricle (via the cerebral aqueduct) and to the central canal of the spinal cord (a narrow, central tube extending the length of the spinal cord). Additional openings in the fourth ventricle allow CSF to flow into the subarachnoid space.

A network of capillaries called the choroid plexus projects into each ventricle. Ependymal cells (a type of neuroglial cell) surround these capillaries. Blood plasma entering the ependymal cells from the capillaries is filtered as it passes into the ventricle, forming CSF. Any material passing from the capillaries to the ventricles of the brain must do so through the ependymal cells because tight junctions linking these cells prevent the passage of plasma between them. Thus, the ependymal cells maintain a blood CSF barrier, controlling the composition of the CSF.

The CSF circulates from the lateral ventricles (where most of the CSF is produced) to the third and then fourth ventricles. From the fourth ventricle, most of the CSF passes into the subarachnoid space, a space within the linings (meninges) of the brain, although some CSF also passes into the central canal of the spinal cord. The CSF returns to the blood through the arachnoid villi located in the dural sinuses of the meninges.

The Meninges

The **meninges** (singular, meninx) are protective coverings of the brain (cranial meninges) and spinal cord (spinal meninges). They consist of three layers of membranous connective tissue:

■ The *dura mater* is the tough outer layer lying just inside the skull and vertebrae. Some characteristics follow:

   ■ In the brain, there are channels within the dura mater, the *dural sinuses,* which contain venous blood returning from the brain to the jugular veins.

   ■ In the spinal cord, the dura mater is often referred to as the *dural sheath.* A fat-filled space between the dura mater and the vertebrae, the epidural space, acts as a protective cushion to the spinal cord.

■ The *arachnoid* (arachnoid mater) is the middle meninx. Projections from the arachnoid, called arachnoid villi, protrude through one layer of the dura mater into the dural sinuses. The arachnoid villi transport the CSF from the subarachnoid space to the dural sinuses. Two cavities border the arachnoid:

- The *subdural space* occurs outside the arachnoid (between the arachnoid and the dura mater).

- The *subarachnoid space* lies inside the arachnoid. This space contains blood vessels and circulates CSF. The fine threads of tissue that spread across this space resemble a spider web and give the arachnoid layer its name (arachnid means spider).

- The *pia mater* is the innermost meninx layer. It tightly covers the brain (following its convolutions) and spinal cord and carries blood vessels that provide nourishment to these nervous tissues.

The Blood-Brain Barrier

Cells in the brain require a very stable environment to ensure controlled and selective stimulation of neurons. As a result, only certain materials are allowed to pass from blood vessels to the brain. Substances such as O_2, glucose, H_2O, CO_2, essential amino acids, and most lipid-soluble substances enter the brain readily. Other substances, such as creatine and urea (wastes transported in the blood), most ions (Na^+, K^+, Cl^-), proteins, and certain toxins either have limited access or are totally blocked from entering the brain. Unfortunately, most antibiotic drugs are equally blocked from entering, while other substances such as caffeine, alcohol, nicotine, and heroin readily enter the brain (because of their lipid solubility). This blood-brain barrier is established by the following:

- Brain capillaries are less permeable than other capillaries because of tight junctions between the endothelial cells in the capillary walls.

- The basal lamina (secreted by the endothelial cells) that surrounds the brain capillaries decreases capillary permeability. This layer is usually absent in capillaries found elsewhere.

- Processes from **astrocytes** (a type of neuroglial cell) cover brain capillaries and are believed to influence capillary permeability in some way.

Cranial Nerves

Cranial nerves are nerves of the PNS that originate from or terminate in the brain. There are 12 pairs of cranial nerves, all of which pass through foramina of the skull. Some cranial nerves are sensory nerves (containing only sensory fibers), some are motor nerves (containing only motor fibers), and some are mixed nerves (containing a combination of sensory and motor nerves). Characteristics of the cranial nerves, which are numbered from anterior to posterior as they attach to the brain, are summarized in Table 11-3 and illustrated in Figure 11-4, earlier in this chapter.

Table 11-3   Characteristics of Cranial Nerves

Cranial Nerve	Nerve Type	Major Functions
I: Olfactory	sensory	smell
II: Optic	sensory	vision
III: Oculomotor	primarily motor	eyeball and eyelid movement; lens shape
IV: Trochlear	primarily motor	eyeball movement; proprioception (superior oblique muscle)
V: Trigeminal: ophthalmic branch	sensory	sensations of touch and pain from facial skin, nose, mouth, teeth, and tongue; proprioception motor control of chewing
V: Trigeminal: maxillary branch	sensory	sensations of touch and pain from facial skin, nose, mouth, teeth, and tongue; proprioception motor control of chewing
V: Trigeminal: mandibular branch	mixed	sensations of touch and pain from facial skin, nose, mouth, teeth, and tongue; proprioception motor control of chewing
VI: Abducens	primarily motor	eyeball movement; proprioception (lateral rectus muscle)
VII: Facial	mixed	movement of facial muscles; tear and saliva secretion; sense of taste and proprioception
VIII: Vestibulocochlear: cochlear branch	sensory	hearing
VIII: Vestibulocochlear: vestibular branch	sensory	sense of equilibrium
IX: Glosso-phayrngeal	mixed	sensations of taste, touch, and pain from tongue and pharynx; chemoreceptors (that monitor O_2 and CO_2); blood pressure receptors; movement of tongue and swallowing; secretion of saliva
X: Vagus	mixed	parasympathetic sensation and motor control of smooth muscles associated with heart, lungs, viscera; secretion of digestive enzymes
XI: Accessory	primarily motor	head movement; swallowing; proprioception
XII: Hypoglossal	primarily motor	tongue movement, speech, and swallowing; proprioception

The Spinal Cord

The spinal cord has two functions:

- *Transmission of nerve impulses.* Neurons in the white matter of the spinal cord transmit sensory signals from peripheral regions to the brain and transmit motor signals from the brain to peripheral regions.

- *Spinal reflexes.* Neurons in the gray matter of the spinal cord integrate incoming sensory information and respond with motor impulses that control muscles (skeletal, smooth, or cardiac) or glands.

The spinal cord is an extension of the brainstem that begins at the foramen magnum and continues down through the vertebral canal to the first lumbar vertebra (L_1). Here, the spinal cord comes to a tapering point, the conus medullaris. The spinal cord is held in position at its inferior end by the filum terminale, an extension of the pia mater that attaches to the coccyx. Along its length, the spinal cord is held within the vertebral canal by denticulate ligaments, lateral extensions of the pia mater that attach to the dural sheath.

The following are external features of the spinal cord (see Figure 11-6):

- Spinal nerves emerge in pairs, one from each side of the spinal cord along its length.

- The cervical nerves form a plexus (a complex interwoven network of nerves—nerves converge and branch).

- The cervical enlargement is a widening in the upper part of the spinal cord (C_4–T_1). Nerves that extend into the upper limbs originate or terminate here.

- The lumbar enlargement is a widening in the lower part of the spinal cord (T_9–T_{12}). Nerves that extend into the lower limbs originate or terminate here.

- The anterior median fissure and the posterior median sulcus are two grooves that run the length of the spinal cord on its anterior and posterior surfaces, respectively.

- The cauda equina are nerves that attach to the end of the spinal cord and continue to run downward before turning laterally to other parts of the body.

- There are four plexus groups: cervical, brachial, lumbar, and sacral. The thoracic nerves do not form a plexus.

Figure 11-6 External features of the spinal cord.

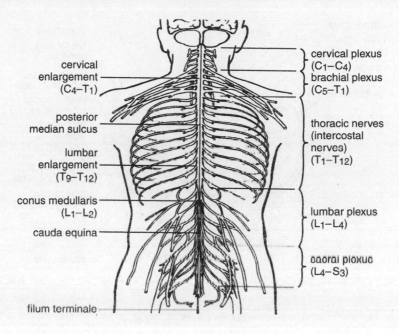

cervical enlargement
(C$_4$–T$_1$)

posterior median sulcus

lumbar enlargement
(T$_9$–T$_{12}$)

conus medullaris
(L$_1$–L$_2$)

cauda equina

filum terminale

cervical plexus
(C$_1$–C$_4$)
brachial plexus
(C$_5$–T$_1$)

thoracic nerves
(intercostal nerves)
(T$_1$–T$_{12}$)

lumbar plexus
(L$_1$–L$_4$)

sacral plexus
(L$_4$–S$_3$)

A cross section of the spinal cord reveals the following features, shown in Figure 11-7:

- Roots are branches of the spinal nerve that connect to the spinal cord. Two major roots form the following:

  - A ventral root (anterior or motor root) is the branch of the nerve that enters the ventral side of the spinal cord. Ventral roots contain motor nerve axons, transmitting nerve impulses from the spinal cord to skeletal muscles.

  - A dorsal root (posterior or sensory root) is the branch of a nerve that enters the dorsal side of the spinal cord. Dorsal roots contain sensory nerve fibers, transmitting nerve impulses from peripheral regions to the spinal cord.

  A dorsal root ganglion is a cluster of cell bodies of a sensory nerve. It is located on the dorsal root.

Figure 11-7 A cross section of the spinal cord.

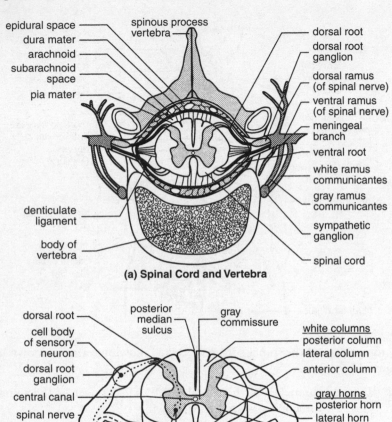

epidural space
dura mater
arachnoid
subarachnoid space
pia mater

spinous process
vertebra

dorsal root
dorsal root ganglion
dorsal ramus (of spinal nerve)
ventral ramus (of spinal nerve)
meningeal branch
ventral root
white ramus communicantes
gray ramus communicantes
sympathetic ganglion
spinal cord

denticulate ligament
body of vertebra

(a) Spinal Cord and Vertebra

dorsal root
cell body of sensory neuron
dorsal root ganglion
central canal
spinal nerve
ventral root
cell body of motor neuron

posterior median sulcus
gray commissure

white columns
posterior column
lateral column
anterior column

gray horns
posterior horn
lateral horn
anterior horn

anterior median fissure

(b) Spinal Cord and Neurons of Spinal Nerve

■ Gray matter appears in the center of the spinal cord in the form of the letter H (or a pair of butterfly wings) when viewed in cross section:

■ The gray commissure is the crossbar of the H.

■ The anterior (ventral) horns are gray matter areas at the front of each side of the H. Cell bodies of motor neurons that stimulate skeletal muscles are located here.

■ The posterior (dorsal) horns are gray matter areas at the rear of each side of the H. These horns contain mostly interneurons that synapse with sensory neurons.

■ The lateral horns are small projections of gray matter at the sides of H. These horns are present only in the thoracic and lumbar regions of the spinal cord. They contain cell bodies of motor neurons in the sympathetic branch of the autonomic nervous system.

■ The central canal is a small hole in the center of the H crossbar. It contains CSF and runs the length of the spinal cord and connects with the fourth ventricle of the brain.

■ White columns (funiculi) refer to six areas of the white matter, three on each side of the H. They are the anterior (ventral) columns, the posterior (dorsal) columns, and the lateral columns.

■ Fasciculi are bundles of nerve tracts within white columns containing neurons with common functions or destinations:

■ Ascending (sensory) tracts transmit sensory information from various parts of the body to the brain.

■ Descending (motor) tracts transmit nerve impulses from the brain to muscles and glands.

Spinal Nerves

There are 31 pairs of spinal nerves (62 total). The following discussion traces a spinal nerve as it emerges from the spinal column (also refer to Figure 11-7):

■ A spinal nerve emerges at two points from the spinal cord, the ventral and dorsal roots.

■ The ventral and dorsal roots merge to form the whole spinal nerve.

■ The spinal nerve emerges from the spinal column through an opening (intervertebral foramen) between adjacent vertebrae. This is true for all spinal nerves except for the first spinal nerve (pair), which emerges between the occipital bone and the atlas (the first vertebra).

- Outside the vertebral column, the nerve divides into the following branches:

  - The dorsal ramus contains nerves that serve the dorsal portions of the trunk.

  - The ventral ramus contains nerves that serve the remaining ventral parts of the trunk and the upper and lower limbs.

  - The meningeal branch reenters the vertebral column and serves the meninges and blood vessels within.

  - The rami communicantes contain autonomic nerves that serve visceral functions.

- Some ventral rami merge with adjacent ventral rami to form a plexus, a network of interconnecting nerves. Nerves emerging from a plexus contain fibers from various spinal nerves, which are then carried together to some target location.

An area of the skin that receives sensory stimuli that pass through a single spinal nerve is called a dermatome. **Dermatomes** are illustrated on a human figure with lines that mark the boundaries of the area where each spinal nerve receives stimuli.

Reflexes

A reflex is a rapid, involuntary response to a stimulus. A reflex arc is the pathway traveled by the nerve impulses during a reflex. Most reflexes are spinal reflexes with pathways that traverse only the spinal cord. During a spinal reflex, information may be transmitted to the brain, but it is the spinal cord, not the brain, that is responsible for the integration of sensory information and a response transmitted to motor neurons. Some reflexes are cranial reflexes with pathways through cranial nerves and the brainstem.

A reflex arc involves the following components, shown in Figure 11-8:

- The receptor is the part of the neuron (usually a dendrite) that detects a stimulus.

- The sensory neuron transmits the impulse to the spinal cord.

- The integration center involves one synapse (monosynaptic reflex arc) or two or more synapses (polysynaptic reflex arc) in the gray matter of the spinal cord. In polysynaptic reflex arcs, one or more interneurons in the gray matter constitute the integration center.

■ A motor neuron transmits a nerve impulse from the spinal cord to a peripheral region.

■ An **effector** is a muscle or gland that receives the impulse from the motor neuron. In somatic reflexes, the effector is skeletal muscle. In autonomic (visceral) reflexes, the effector is smooth or cardiac muscle, or a gland.

Figure 11-8 A reflex arc.

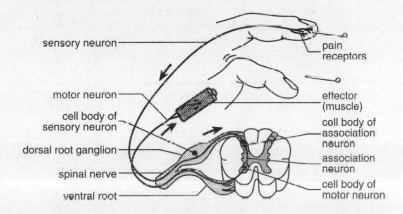

Some examples of reflexes follow:

■ A stretch reflex is a monosynaptic reflex that is a response to a muscle that has been stretched (the knee-jerk reflex is an example). When receptors in muscles, called muscle spindles, detect changes in muscle length, they stimulate, through a reflex arc, the contraction of a muscle. Stretch reflexes help maintain posture by stimulating muscles to regain normal body position.

■ A flexor (withdrawal) reflex is a polysynaptic reflex that causes a limb to be withdrawn when it encounters pain (refer to Figure 11-8).

■ A monosynaptic reflex is, typically, a reflex that does not involve the brain. See Figure 11-9. There is no association neuron in the spinal cord; therefore, information does not go to the brain. An example of a monosynaptic reflex is the patellar reflex, sometimes called the knee-jerk reflex.

Figure 11-9 The patellar reflex.

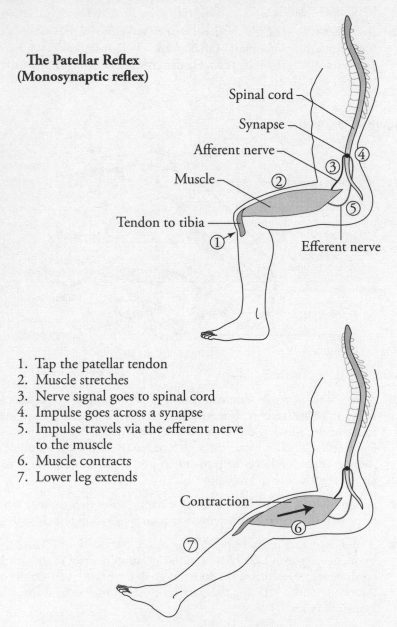

**The Patellar Reflex
(Monosynaptic reflex)**

Spinal cord

Synapse

Afferent nerve

Muscle

Tendon to tibia

Efferent nerve

1. Tap the patellar tendon
2. Muscle stretches
3. Nerve signal goes to spinal cord
4. Impulse goes across a synapse
5. Impulse travels via the efferent nerve to the muscle
6. Muscle contracts
7. Lower leg extends

Contraction

The Autonomic Nervous System

The peripheral nervous system consists of the **somatic nervous system (SNS)** and the **autonomic nervous system (ANS).** The SNS consists of motor neurons that stimulate skeletal muscles. In contrast, the ANS consists of motor neurons that control smooth muscles, cardiac muscles, and glands. In addition, the ANS monitors visceral organs and blood vessels with sensory neurons, which provide input information for the CNS.

The ANS is further divided into the sympathetic nervous system and the parasympathetic nervous system. Both of these systems can stimulate and inhibit effectors. However, the two systems work in opposition—where one system stimulates an organ, the other inhibits. Working in this fashion, each system prepares the body for a different kind of situation, as follows:

■ The *sympathetic nervous system* prepares the body for situations requiring alertness or strength, or situations that arouse fear, anger, excitement, or embarrassment ("fight-or-flight" situations). In these kinds of situations, the sympathetic nervous system stimulates cardiac muscles to increase the heart rate, causes dilation of the bronchioles of the lungs (increasing oxygen intake), and causes dilation of blood vessels that supply the heart and skeletal muscles (increasing blood supply). The adrenal medulla is stimulated to release epinephrine (adrenalin) and norepinephrine (noradrenalin), which in turn increases the metabolic rate of cells and stimulates the liver to release glucose into the blood. Sweat glands are stimulated to produce sweat. In addition, the sympathetic nervous system reduces the activity of various "tranquil" body functions, such as digestion and kidney functioning.

■ The *parasympathetic nervous system* is active during periods of digestion and rest. It stimulates the production of digestive enzymes and stimulates the processes of digestion, urination, and defecation. It reduces blood pressure and heart and respiratory rates and conserves energy through relaxation and rest.

In the SNS, a single motor neuron connects the CNS to its target skeletal muscle. In the ANS, the connection between the CNS and its effector consists of two neurons—the preganglionic neuron and the postganglionic neuron. The synapse between these two neurons lies outside the CNS, in an autonomic ganglion. The axon (preganglionic axon) of a preganglionic neuron enters the ganglion and forms a synapse with the dendrites of the postganglionic neuron. The axon of the postganglionic neuron emerges from the ganglion and travels to the target organ (see Figure 11-10). There are three kinds of autonomic ganglia:

■ The sympathetic trunk, or chain, contains sympathetic ganglia called paravertebral ganglia. There are two trunks, one on either side of the vertebral column along its entire length. Each trunk consists of ganglia connected by fibers, like a string of beads.

■ The prevertebral (collateral) ganglia also consist of sympathetic ganglia. Preganglionic sympathetic fibers that pass through the sympathetic trunk (without forming a synapse with a postganglionic neuron) synapse here. Prevertebral ganglia lie near the large abdominal arteries, which the preganglionic fibers target.

■ Terminal (intramural) ganglia receive parasympathetic fibers. These ganglia occur near or within the target organ of the respective postganglionic fiber.

Figure 11-10   The target organs of the different nervous systems.

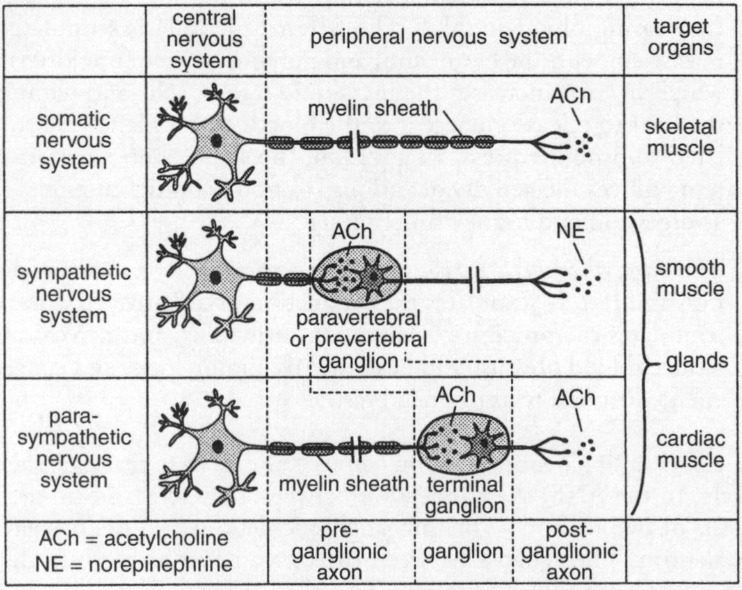

	central nervous system	peripheral nervous system		target organs
somatic nervous system		myelin sheath / ACh		skeletal muscle
sympathetic nervous system		ACh — paravertebral or prevertebral ganglion — NE		smooth muscle / glands
para-sympathetic nervous system		myelin sheath — terminal ganglion — ACh — ACh		cardiac muscle
ACh = acetylcholine NE = norepinephrine	pre-ganglionic axon	ganglion	post-ganglionic axon	

A comparison of the sympathetic and parasympathetic pathways follows (see Figure 11-11):

■ *Sympathetic nervous system.* Cell bodies of the preganglionic neurons occur in the lateral horns of gray matter of the 12 thoracic and first 2 lumbar segments of the spinal cord. (For this reason, the sympathetic system is also called the thoracolumbar division.) Preganglionic fibers leave the spinal cord within spinal nerves through the ventral roots (together with the PNS motor neurons). The preganglionic fibers then branch away from the nerve through white rami (white rami communicantes) that connect with the sympathetic trunk. White rami are white because they contain myelinated fibers. A preganglionic fiber that enters the trunk may synapse in the first ganglion it enters, travel up or down the trunk to synapse with another ganglion, or pass through the trunk and synapse outside the trunk. Postganglionic fibers that originate in ganglia within the sympathetic trunk leave the trunk through gray rami (gray rami communicantes) and return to the spinal nerve, which is followed until it reaches its target organ. Gray rami are gray because they contain unmyelinated fibers.

■ *Parasympathetic nervous system.* Cell bodies of the preganglionic neurons occur in the gray matter of sacral segments S_2–S_4 and in the brainstem (with motor neurons of their associated cranial nerves III, VII, IX, and X). (For this reason, the parasympathetic system is also called the craniosacral division, and the fibers arising from this division are called the cranial outflow or the sacral outflow, depending on their origin.) Preganglionic fibers of the cranial outflow accompany the PNS motor neurons of cranial nerves and have terminal ganglia that lie near the target organ. Preganglionic fibers of the sacral outflow accompany the PNS motor neurons of spinal nerves. These nerves emerge through the ventral roots of the spinal cord and have terminal ganglia that lie near the target organ.

Figure 11-11   A comparison of the sympathetic and parasympathetic pathways.

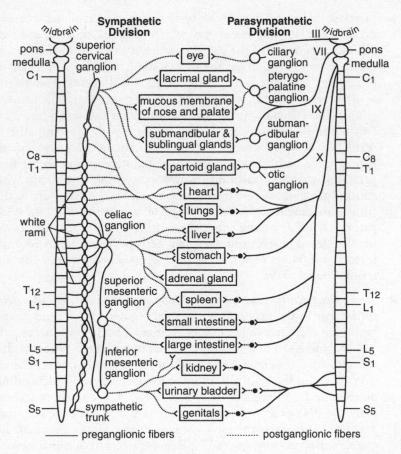

Chapter Check-Out

Q&A

1. Which branch of the efferent nervous system controls the activities of voluntary movement?
2. True or False: The sympathetic nervous system can simultaneously cause certain blood vessels to contract while others are stimulated to dilate.

3. Which of the following statement(s) is/are *true* concerning the diencephalon?
   a. forms a portion of the brainstem
   b. contains the superior colliculus
   c. primarily composed of the hypothalamus and thalamus
   d. derived from the prosencephalon
   e. located inferior to the cerebellum

4. True or False: The dorsal root carries afferent nerve fibers.

5. The _____ _____ is the tapered point of the spinal cord at the level of the second lumbar vertebra.

Answers: 1. efferent (or motor); somatic may also be a possible answer, **2.** T (only the SNS regulates blood flow), **3.** a, c, and d, **4.** T, **5.** conus medullaris

Chapter 12

THE SENSORY SYSTEM

Chapter Check-In

❑ Finding out how sensory receptors are classified

❑ Understanding the different types of sensory stimuli that the body detects

❑ Naming the components of the eye and its accessory organs that aid in vision

❑ Describing the major regions of the ear and the process of hearing

❑ Appreciating the other special senses, such as taste, smell, and equilibrium

One of the characteristics of a living organism is its ability to respond to stimuli. The human sensory system is highly evolved and processes thousands of incoming messages simultaneously. This complexity allows you to be aware of your surroundings and take appropriate actions.

Sensory Receptors

Sensory receptors are dendrites of sensory neurons specialized for receiving specific kinds of stimuli. Sensory receptors are classified by three methods:

■ Classification by receptor complexity:

   ■ Free nerve endings are dendrites whose terminal ends have little or no physical specialization.

   ■ Encapsulated nerve endings are dendrites whose terminal ends are enclosed in a capsule of connective tissue.

- Sense organs (such as the eyes and ears) consist of sensory neurons with receptors for the special senses (vision, hearing, smell, taste, and equilibrium) together with connective, epithelial, or other tissues.

- Classification by location:

  - **Exteroceptors** occur at or near the surface of the skin and are sensitive to stimuli occurring outside or on the surface of the body. These receptors include those for tactile sensations, such as touch, pain, and temperature, as well as those for vision, hearing, smell, and taste.

  - **Interoceptors** (visceroceptors) respond to stimuli occurring in the body from visceral organs and blood vessels. These receptors are the sensory neurons associated with the autonomic nervous system.

  - **Proprioceptors** respond to stimuli occurring in skeletal muscles, tendons, ligaments, and joints. These receptors collect information concerning body position and the physical conditions of these locations.

- Classification by type of stimulus detected:

  - Mechanoreceptors respond to physical force such as pressure (touch or blood pressure) and stretch.

  - Photoreceptors respond to light.

  - Thermoreceptors respond to temperature changes.

  - Chemoreceptors respond to dissolved chemicals during sensations of taste and smell and to changes in internal body chemistry such as variations of O_2, CO_2, or H^+ in the blood.

  - Nociceptors respond to a variety of stimuli associated with tissue damage. The brain interprets the pain.

The Somatic Senses

The somatic (general) senses collect information about cutaneous sensations (tactile sensations on the surface of the skin) and proprioceptive sensations. The following stimuli are detected:

■ Tactile stimuli are detected by mechanoreceptors and produce sensations of touch and pressure:

  ■ Merkel discs are receptors with free nerve endings that detect surface pressure (light touch). They are located deep in the epidermis.

  ■ Root hair plexuses are receptors with free nerve endings that surround hair follicles and detect hair movement.

  ■ Corpuscles of touch (Meissner's corpuscles) are receptors with encapsulated nerve endings located in the dermal paillae (near the surface) of the skin that detect surface pressure (light touch).

  ■ Pacinian corpuscles are encapsulated nerve receptors that detect deep pressure and are located in the subcutaneous layer (below the skin).

■ Thermal stimuli are detected by free nerve ending thermoreceptors sensitive to heat or cold.

■ Pain stimuli are detected by free nerve ending nociceptors.

■ Proprioceptive stimuli are detected by the following receptors:

  ■ Muscle spindles are mechanoreceptors located in skeletal muscles. They consist of specialized skeletal muscle fibers enclosed in a spindle-shaped capsule made of connective tissue.

  ■ Golgi tendon organs are mechanoreceptors located at the junctions of tendons and muscles.

  ■ Joint kinesthetic receptors are mechanoreceptors located in synovial joints.

Vision

The eye is supported by the following accessory organs:

■ The eyebrows shade the eyes and help keep perspiration that accumulates on the forehead from running into the eyes.

■ The eyelids (palpebrae) lubricate, protect, and shade the eyeballs. Contraction of the levator palpebrae superioris muscle raises the upper eyelid. Each eyelid is supported internally by a layer of connective tissue, the tarsal plate. Tarsal (Meibomian) glands embedded in the tarsal plate produce secretions that prevent the upper and lower eyelids from sticking together. The inner lining of the eyelid, the conjunctiva, is a mucous membrane that produces secretions that lubricate the eyeball. The conjunctiva continues beyond the eyelid, folding back to cover the white of the eye.

■ The eyelashes, on the borders of the eyelids, help protect the eyeball. Nerve endings at the base of the hairs initiate a reflex action that closes the eyelids when the eyelashes are disturbed.

■ The lacrimal apparatus produces and drains tears. Tears (lacrimal fluid) are produced by the lacrimal glands, which lie above each eye (toward the lateral edge). In each eye, tears flow across the eyeball and enter two openings (lacrimal puncta) into lacrimal canals that lead to the lacrimal sac. From here, the tears drain through the nasolacrimal duct into the nasal cavity. Tears contain antibodies and lysozyme (a bacteria-destroying enzyme).

■ Six extrinsic eye muscles provide fine motor control for the eyeballs. These are the lateral, medial, superior, and inferior rectus muscles and the inferior and superior oblique muscles.

The eyeball is a hollow sphere whose wall consists of three tunics (layers), shown in Figure 12-1.

The three tunics of the eye are described below:

■ *Fibrous tunic:* The outer fibrous tunic consists of avascular connective tissue called the sclera. The forward ⅙ portion of this tunic is the cornea, a transparent layer of collagen fibers that forms a window for entering light. The remainder of the fibrous tunic is the sclera. Consisting of tough connective tissue, the sclera maintains the shape of the eyeball and provides for the attachment of the eye muscles. The visible forward portion of the sclera is the white of the eye.

■ *Vascular tunic:* The middle vascular tunic (uvea) consists of three highly vascularized (as the name implies), pigmented parts (the iris, the ciliary body, and the choroid):

■ The iris is the colored portion of the eye that opens and closes to control the size of its circular opening, the pupil. The size of the pupil regulates the amount of light entering the eye and helps bring objects into focus.

Figure 12-1 Details of the eye and the retina.

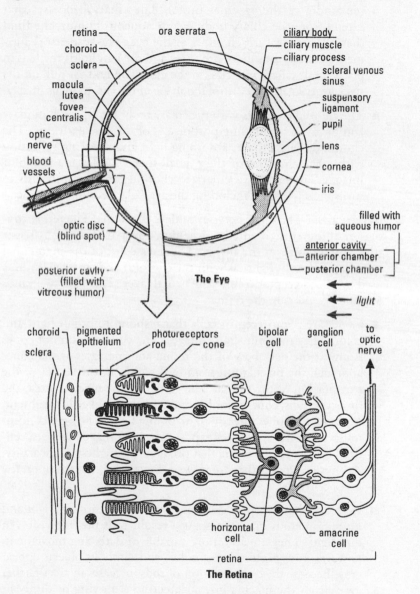

The Eye

The Retina

- The ciliary body lies between the iris and the choroids (the remainder of the vascular tunic). The ciliary processes that extend from the ciliary body secrete aqueous humor, the fluid that fills the forward chamber of the eye. The suspensory ligament between the ciliary processes and the lens holds the lens in place, while ciliary muscles (in the ciliary body) that pull on the suspensory ligament control the shape of the lens to focus images.

- The choroid connects with the ciliary body at a jagged boundary and forms the remaining portion (⅚) of the vascular tunic. The choroid is dark brown, absorbing light and reducing reflection within the chamber of the eyeball that would otherwise blur images. The highly vascularized choroid provides nutrients to surrounding tissues, including the avascularized fibrous tunic.

- *Nervous tunic:* The inner nervous tunic is the retina. The retina consists of an outer pigmented epithelium covered by nervous tissue (the neural layer) on the inside. The dark color of the pigmented epithelium absorbs light (as with the choroid) and stores vitamin A used by photoreceptor cells in the neural layer. There are two kinds of photoreceptors in the retina:

  - Cones are photoreceptor cells that respond to bright light and color. They transmit sharp images. The concentration of cones is low at the periphery of the retina and increases as the cones approach the macula lutea, an oval region in the center of the rear portion of the retina. The center of the macula lutea, the fovea centralis, contains only cones; other retinal cells are absent, exposing the cones directly to incoming light. The high concentration of cones and direct exposure to light make the fovea centralis the site on the retina that provides the highest visual acuity. As a result, images that are viewed directly are focused upon the fovea centralis.

  - Rods are photoreceptor cells that are more sensitive to light and more numerous than cones. As a result, rods provide vision in dim light. They are also more capable of detecting movement; however, rods cannot detect color, and dimly lit objects appear gray. Because the concentration of rods increases in areas farther away from the macula lutea, detecting a moving or dimly lit object can be more effectively achieved by looking slightly away from the object.

Within the nervous tunic, photoreceptor cells (rods and cones) form synapses with other nerve cells (refer to Figure 12-1). When stimulated by light, rods and cones pass graded potentials to bipolar cells, which in turn pass graded potentials to the ganglion cells. The graded potentials may be modified by horizontal cells and amacrine cells that link adjacent photoreceptors or ganglion cells, respectively. Action potentials are ultimately generated by ganglion cells. The axons of all the ganglion cells gather at the optic disc and exit the nervous tunic through the optic disc as the optic nerve. The optic disc is a blind spot because photoreceptors are absent here.

The lens of the eye consists of tightly packed cells arranged in successive layers (as in an onion) and filled with transparent proteins called crystallins. The lens divides the interior of the eyeball into two cavities:

- The anterior cavity, the area in front of the lens, is subdivided by the iris and ciliary body into the anterior chamber and the posterior chamber. Capillaries in the ciliary body produce a clear fluid, the aqueous humor, which flows into the posterior chamber, through the pupil, and into the anterior chamber. The aqueous humor then drains into veins through a channel (scleral venous sinus, or canal of Schlemm) that encircles the eye where the cornea and sclera join. The aqueous humor, which is continuously replaced, provides pressure to maintain the shape of the forward portion of the eye and supplies O_2 and nutrients to the avascular lens and cornea.

- The posterior cavity, the area behind the lens, is filled with a clear gel, the vitreous humor. The vitreous humor, which is produced during embryonic development and is not replaced, holds the lens and retina in position and maintains the shape of the eye.

The process of sight involves converting light energy to chemical energy. Features of the process follow:

- The outer segments of rods and cones contain numerous folds that increase the surface area exposed to light.

- Photopigments (visual pigments) in the outer segments respond to light by changing their chemical structure. Each visual photopigment consists of two parts: retinal (a vitamin A derivative) and opsin (a glycoprotein). There are four different kinds of photopigments because the opsin in each has a slightly different structure, enabling each photopigment to absorb a different range of light wavelengths (different colors).

- There are three kinds of cones, each possessing a different photopigment, and each sensitive to a different range of light wavelengths. A mixed stimulation of the three different cones (called red, green, and blue for their optimally absorbed wavelengths) provides for the perception of varied colors.

- There is only one kind of rod, with a single kind of photopigment, rhodopsin.

- When a photopigment absorbs light, retinal changes shape, causing it to separate from opsin. Because the product is colorless, the process is called bleaching.

- Unlike most other neurons, photoreceptors continually secrete a neurotransmitter when in the resting (unstimulated), polarized condition. When a photoreceptor is stimulated, the freed opsin becomes chemically active and initiates a series of chemical reactions that close the Na^+ channels in the plasma membrane. Because the Na^+/K^+ pump continues to pump Na^+ out of the cell, the plasma membrane becomes hyperpolarized. Hyperpolarization stops the normal secretion of the neurotransmitter, which in turn stimulates a graded potential in the bipolar cells.

- Photopigments are regenerated when opsin is enzymatically reattached to retinal. In bright light, the very light-sensitive rhodopsin in rods cannot be regenerated as fast as it is broken down, so most of the rhodopsin remains inactive. As a result, the less light-sensitive pigments in cones are active in normal daylight. When eyes move from bright conditions to dark conditions, the photopigments in cones are too insensitive to detect light, and the rhodopsin in rods is still bleached from its exposure to bright light. Vision returns as rhodopsin is regenerated, a process called dark adaptation. When eyes move from dark to bright conditions, very light-sensitive rods are suddenly overwhelmed with stimulation, producing the sensation of glare. Light adaptation occurs as the rhodopsin in rods is completely bleached and the less light-sensitive cones resume activity.

When light reflected from an object enters the eye, the following processes occur:

- *Light refraction:* When light rays pass from one substance to another substance of different density, the rays bend, or refract. The amount of bending depends on the angle of incidence of the light ray and the degree to which the densities of the two substances differ. When distant objects are sighted, the normal curvature of the lens appropriately compensates for the refraction due to the differences in densities among the aqueous humor, the lens, and the vitreous humor.

■ *Lens accommodation:* Light rays from near objects enter the eyeball at more divergent angles than rays from distant objects. Thus, when near objects are sighted, muscles pull on the lens to increase its curvature so that the more divergent rays of the close object are properly refracted upon the retina.

■ *Pupil constriction:* One function of the pupil is to regulate the amount of light that enters the posterior cavity so that the retina receives the appropriate amount of stimulation. In addition, when near objects are sighted, the pupil constricts (accommodation pupillary reflex) to block the most divergent light rays that cannot be brought into focus by the accommodation of the lens. Reading under low levels of light may be difficult because the pupils are dilated to allow all available light to enter rather than constricted to improve focusing. Increasing the amount of light improves the ability to read because the surplus light permits the pupils to constrict and the lens to focus a more narrow beam of light rays.

■ *Eyeball convergence:* Both eyes point in the same direction when viewing a distant object. When near objects are sighted, the eyes must be directed medially to simultaneously view the object, a process called convergence.

Nerve impulses generated by visual stimuli travel along the axons of ganglion cells within the two optic nerves. Before entering the brain, axons representing the medial portions of the visual fields of each eye cross over at the optic chiasma. After the crossover, the axons, now forming the optic tract, enter the thalamus. Processed visual stimuli are then carried to visual areas of the occipital lobes of both cerebral hemispheres by nerve pathways called the optic radiations. Because of the partial crossover at the optic chiasma, each cerebral hemisphere receives the lateral portion of the visual field of the eye on the same side of the body as the cerebral hemisphere, but the medial portion of the visual field of the eye is interpreted on the opposite side of the body. In addition, because of the action of the lens, the image that forms on the retina and that is sent to the brain is inverted and reversed right to left. The brain, however, interprets all of this seemingly disparate visual information into a coherent perception of the real world.

Hearing

The organ of hearing, the ear, consists of three major regions, shown in Figure 12-2.

- The outer (external) ear consists of the auricle (pinna), a flap of elastic cartilage that protrudes from the head, and the external auditory canal (meatus), a tube that enters the temporal bone. The canal is lined with ceruminous glands that secrete cerumen (earwax), a sticky substance that traps dirt and other foreign objects. The eardrum (tympanic membrane), at the internal end of the external auditory canal, vibrates in response to incident sound waves.

- The middle ear (tympanic cavity) is an air-filled cavity within the temporal bone. It contains three small bones, the auditory ossicles. These bones, called the malleus, incus, and stapes, act as a lever system that amplifies and transfers vibrations of the eardrum to the inner ear. The malleus at one end connects to the eardrum, while the stapes at the other end attaches with ligaments to the oval window, a small, membrane-covered opening into the inner ear. Synovial joints connect the incus, the center bone of the auditory ossicles, to the malleus and stapes on each side. A second membrane-covered opening to the inner ear, the round window (secondary tympanic membrane), lies just below the oval window. A third opening leads to the auditory (Eustachian) tube, which connects the middle ear to the upper throat. The auditory tube allows pressure differences between the middle and outer ear to equalize, thus reducing tension on the eardrum. Two muscles in the middle ear, the tensor tympani and the stapedius, connect to the malleus and stapes, respectively. Contraction of these two muscles restricts the movement of the eardrum and auditory ossicles, reducing damage that may occur when they are exposed to excessive vibration from loud noises.

- The inner (internal) ear, also called the labyrinth, is a system of double-walled canals. The canals consist of an outer bony (osseous) labyrinth that encloses an inner membranous labyrinth. Perilymph fills the space between the two labyrinths, and endolymph fills the inner labyrinth. This double-layer labyrinth structure is found throughout the following inner ear structures. This labyrinth is made of three semicircular canals and a snail-shaped cochlea (see Figure 12-2).

Figure 12-2 The three major regions of the ear are the outer ear, the middle ear, and the inner ear.

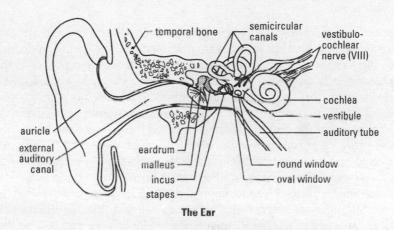

The Ear

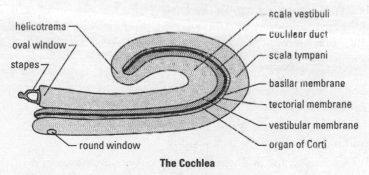

The Cochlea

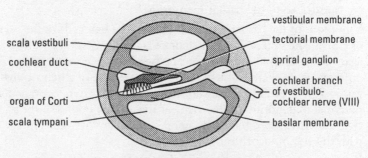

The Cochlea (Cross Section)

Three semicircular canals contain receptor cells for determining angular movements of the head. This information is used for establishing equilibrium.

The cochlea is a coiled canal that contains receptor cells that respond to vibrations transferred from the middle ear. The interior of the cochlea is divided into three regions, or scalas: the scala vestibuli, the scala tympani, and the cochlear duct (scala media). The scalas are tubular channels that follow the coiled curvature of the cochlea. At the middle ear, the scala vestibuli and the scala tympani connect to the oval and round windows, respectively. At the other end of the cochlea, in a region called the **helicotrema,** these two scalas join, allowing free movement of the perilymph within. The third scala, the cochlear duct, is separated from the scala vestibuli and the scala tympani by the vestibular membrane and the basilar membrane, respectively. The cochlear duct is filled with endolymph and internally lined with the organ of Corti. The organ of Corti contains numerous hair cells. The bases of the hair cells are attached to the basilar membrane, while hairlike microvilli called stereocilia project upward into an overlying gel, the tectorial membrane. The stereocilia are receptors for vibrations that are produced when the underlying basal membrane moves relative to the overlying tectorial membrane.

The process of hearing occurs as follows:

1. Sound waves, funneled into the outer ear by the auricle, cause the eardrum to vibrate.

2. Vibrations of the eardrum are amplified and transferred by the auditory ossicles of the oval window.

3. Vibrations on the oval window produce pressure waves in the perilymph of the scala vestibuli and the scala tympani. These vibrations are transferred to the basilar membrane.

4. Vibrations of the basilar membrane move the hair cells of the organ of Corti. The stereocilia of the hair cells bend when they move against the tectorial membrane. The bending generates a graded potential in the hair cell, which causes the release of a neurotransmitter at its base. The neurotransmitter in turn generates an action potential in dendrites of the cochlear nerve. Cell bodies of the cochlear nerve assemble in the spiral ganglia, and its axons merge with the vestibulocochlear nerve.

5. Pressure waves in the perilymph of the scala tympani cause the round window to bulge into the middle ear. This allows vibrational movements of the perilymph (and indirectly the endolymph) that, as an incompressible fluid, would not otherwise be able to vibrate within the surrounding rigid temporal bone.

Equilibrium

The vestibule lies between the semicircular canals and the cochlea. It contains two bulblike sacs, the saccule and utricle, whose membranes are continuous with those of the cochlea and semicircular canals, respectively. The saccule and utricle contain receptors that help maintain equilibrium.

Equilibrium is maintained in response to two kinds of motion:

■ Static equilibrium maintains the position of the head in response to linear movements of the body, such as starting to walk or stopping.

■ Dynamic equilibrium maintains the position of the head in response to rotational motion of the body, such as rocking (as in a boat) or turning.

The perception of equilibrium occurs in the vestibular apparatus. Motion in the following two structures is detected as follows:

■ The vestibule is the primary detector of changes in static equilibrium. A sensory receptor called a macula is located in the walls of the saccule and utricle, the two bulblike sacs of the vestibule. A macula contains numerous receptor cells called hair cells, from which numerous stereocilia (long microvilli) and a single kinocilium (a true cilium) extend into a glycoprotein gel, the otolithic membrane. Calcium carbonate crystals called otoliths pervade the otolithic membrane, increasing its density and thus its responsiveness to changes in motion. Changes in linear motion cause the otolithic membrane to move forward and backward in the utricle or up and down in the saccule. The movement of the otolithic membrane causes similar movements in the embedded stereocilia of the hair cells, which in turn initiate graded potentials.

■ The semicircular canals are the primary detector of changes in dynamic equilibrium. The three canals, individually called the anterior, posterior, and lateral canals, are arranged at right angles to one another. The expanded base of each canal, called an **ampulla,** contains a sensory receptor, or crista ampullaris. Like the maculae of the vestibule, each crista ampullaris contains numerous hair cells whose stereocilia and kinocilium protrude into a gelatinous matrix, the cupula (which is analogous to the otolithic membranes of the maculae). Changes in rotational motion cause the cupula and the embedded stereocilia to move, which stimulates the hair cells to generate a graded potential.

Graded potentials in the hair cells of the maculae and cristae result in changes in the amounts of neurotransmitter secreted. In response to these changes, action potentials are generated in the fibers of the vestibular nerve, which subsequently joins the vestibulocochlear nerve. From here, the nerve impulses travel to the pons and the medulla oblongata.

Smell

The sense of smell, or olfactory sense, occurs in olfactory epithelium that occupies a small area on the roof of the nasal cavity. The olfactory receptor cells are bipolar neurons whose dendrites have terminal knobs with hairlike cilia protruding beyond the epithelial surface. The cilia, or olfactory hairs, initiate an action potential when they react with a molecule from an inhaled vapor. However, molecules of the vapor must first dissolve in the mucus that covers the cilia before they can be detected. The action potential is transmitted along the axons of the olfactory receptor cells (which form the olfactory nerves) to the olfactory bulbs, where they synapse with sensory neurons of the olfactory tract.

Other cells of the olfactory epithelium include columnar supporting cells and basal cells. The basal cells continually divide to produce new olfactory receptor cells that, because of their short life, need regular replacement. The replacement of olfactory receptor cells is unusual because most other nerve cells cannot be replaced.

The mucus that lines the olfactory epithelium is produced by olfactory glands that occupy the connective tissue above the olfactory epithelium.

Taste

The sense of taste, or gustatory sense, occurs in the taste buds. Located primarily on the tongue, taste buds reside in papillae, the bumps on the tongue that give it a rough texture. The taste bud consists of supporting cells, basal cells, and gustatory (taste) receptor cells arranged in the shape of a glove with an opening, or taste pore, to the outside located at the top. A long microvilli, or gustatory hair, from each gustatory receptor cell within the taste bud projects through the taste pore. Gustatory hairs generate action potentials when stimulated by chemicals that are dissolved in the saliva.

Basal cells are actively dividing epithelial cells. The daughter cells of basal cells develop into supporting cells, which subsequently mature into gustatory receptor cells. Because they are easily damaged by the activities that occur in the mouth, gustatory receptor cells are short-lived and replaced about every ten days.

The various gustatory cells respond to numerous taste sensations such as sweet, bitter, sour, salty, and umami (the taste of some amino acids and aged cheese). Research is determining that there are possibly other taste sensations. All tastes arise from a mixture of the stimulated taste receptors in combination with olfactory sensations. Taste buds on certain areas of the tongue seem to specialize in certain tastes. For example, the sensation of sweetness is best detected at the front of the tongue, while bitterness is best detected at the back of the tongue.

Chapter Check-Out

Q&A

1. True or False: The cochlear duct (or scala media) is filled with perilymph.

2. Which of the following statement(s) is/are true regarding vision?

   a. Photopigments in the outer segments alter their structure in response to light.

   b. Three kinds of rods are within the retina.

   c. Cones provide vision in dim light.

   d. During accommodation, the pupil constricts when near objects are sighted.

3. True or False: Proprioception is realizing the position in which the body currently resides.

4. _____ equilibrium maintains the position of the head in response to rotational motion of the body.

Answers: 1. F, **2.** a, d, **3.** T, **4.** Dynamic

Chapter 13

THE ENDOCRINE SYSTEM

Chapter Check-In

❑ Classifying the four different types of hormones

❑ Describing how hormones affect their target organs and help regulate homeostasis within the body

❑ Listing the hormones within the body and their functions

Although It may seem like a far-fetched notion that a small chemical can enter the bloodstream and cause an action at a distant location in the body, this scenario occurs every day. The ability to maintain homeostasis and respond to stimuli is largely due to hormones secreted within your body. Without hormones, you could not grow, maintain a constant temperature, produce offspring, or perform the basic actions that are essential for life.

Hormones

The endocrine system produces hormones that are instrumental in maintaining homeostasis and regulating reproduction and development. A **hormone** is a chemical messenger produced by a cell that effects specific change in the cellular activity of other cells (target cells). Unlike exocrine glands (which produce substances such as saliva, milk, stomach acid, and digestive enzymes), endocrine glands do not secrete substances into ducts (tubes). Instead, endocrine glands secrete their hormones directly into the surrounding extracellular space. The hormones then diffuse into nearby capillaries and are transported throughout the body in the blood.

The endocrine and nervous systems often work toward the same goal—both influence other cells with chemicals (hormones and neurotransmitters). However, they attain their goals differently. **Neurotransmitters** act immediately (within milliseconds) on adjacent muscles, glands, or other nervous cells, and their effect is short-lived. In contrast, hormones take longer to produce their intended effect (anywhere from seconds to days); may affect any cell, nearby or distant; and produce effects that last as long as they remain in the blood (up to several hours).

Hormones can be chemically classified into four groups:

■ Amino acid-derived hormones are modified amino acids.

■ Polypeptide and protein hormones are chains of amino acids of less than or more than about 100 amino acids, respectively. Some protein hormones are actually glycoproteins, containing glucose or other carbohydrate groups.

■ Steroid hormones are lipids that are synthesized from cholesterol. Steroids are characterized by four interlocking carbohydrate rings.

■ Eicosanoids are lipids that are synthesized from the fatty acid chains of phospholipids found in the plasma membrane.

Mechanisms of hormone action

Hormones circulating in the blood diffuse into the interstitial fluids surrounding the cell. Cells with specific receptors for a hormone respond with an action that is appropriate for the cell. Because of the specificity of hormone and target cell, the effects produced by a single hormone may vary among different kinds of target cells.

Hormones activate target cells by one of two methods, depending on the chemical nature of the hormone:

■ Lipid-soluble hormones (steroid hormones and hormones of the thyroid gland) diffuse through the cell membranes of target cells. The lipid-soluble hormone then binds to a receptor protein that in turn activates a DNA segment that turns on specific genes. The proteins produced as a result of the transcription of the genes and subsequent translation of mRNA act as enzymes that regulate specific physiological cell activity.

■ Water-soluble hormones (polypeptide, protein, and most amino acid hormones) bind to a receptor protein on the plasma membrane of the cell. The receptor protein in turn stimulates the production of one of the following second messengers:

- Cyclic AMP (cAMP) is produced when the receptor protein activates another membrane-bound protein called a G protein. The G protein activates adenylate cyclase, the enzyme that catalyzes the production of cAMP from ATP. Cyclic AMP then triggers an enzyme that generates specific cellular changes.

- Inositol triphosphate (IP_3) is produced from membrane phospholipids. IP_3 in turn triggers the release of Ca^{2+} from the endoplasmic reticulum, which then activates enzymes that generate cellular changes.

Control of hormone production

Endocrine glands release hormones in response to one (or more) of the following stimuli:

- Hormones from other endocrine glands

- Chemical characteristics of the blood (other than hormones)

- Neural stimulation

Most hormone production is regulated by a **negative feedback** system. The nervous system and certain endocrine tissues monitor various internal conditions of the body. If action is necessary to maintain homeostasis, hormones are released, either directly by an endocrine gland or indirectly via the action of the hypothalamus of the brain, which stimulates other endocrine glands to release hormones. The hormones activate target cells, which initiate physiological changes that adjust body conditions. When normal conditions have been restored, the corrective action (the production of hormones) is discontinued. Thus, in negative feedback, when the original (abnormal) condition has been repaired, or negated, corrective actions decrease or are discontinued. For example, the amount of glucose in the blood regulates the secretion of insulin and glucagons through negative feedback.

The production of some hormones is regulated by **positive feedback.** In such a system, hormones cause a condition to intensify (rather than decrease). As the condition intensifies, hormone production increases. Such positive feedback is uncommon but does occur during childbirth (hormone levels build with increasingly intense labor contractions) and lactation (where hormone levels increase in response to nursing, which causes milk production to increase).

The Hypothalamus and Pituitary Glands

The hypothalamus makes up the lower region of the diencephalon and lies just above the brain stem. The pituitary gland (hypophysis) is attached to the bottom of the hypothalamus by a slender stalk called the infundibulum. The pituitary gland consists of two major regions: the anterior pituitary gland (anterior lobe or adenohypophysis) and the posterior pituitary gland (posterior lobe or neurohypophysis).

The hypothalamus oversees many internal body conditions. It receives nervous stimuli from receptors throughout the body and monitors chemical and physical characteristics of the blood, including temperature; blood pressure; and nutrient, hormone, and water content. When deviations from homeostasis occur or when certain developmental changes are required, the hypothalamus stimulates cellular activity in various parts of the body by directing the release of hormones from the anterior and posterior pituitary glands. The hypothalamus communicates directives with these glands by one of the following two pathways:

■ Communication between the hypothalamus and the anterior pituitary occurs through chemicals (releasing hormones and inhibiting hormones) that are produced by the hypothalamus and delivered to the anterior pituitary through blood vessels in the infundibulum. The releasing and inhibiting hormones are produced by specialized neurons of the hypothalamus, called neurosecretory cells. The hormones are released into a capillary network (primary plexus) and transported through veins (hypophyseal portal veins) to a second capillary network (secondary plexus) that supplies the anterior pituitary. The primary plexus and the hypophyseal portal veins are in the infundibulum and the secondary plexus is in the anterior pituitary. The hormones then diffuse from the secondary plexus into the cells of the anterior pituitary, where they initiate the production of specific hormones by the anterior pituitary. The releasing and inhibiting hormones secreted by the hypothalamus and the hormones produced in response by the anterior pituitary are listed in Table 13-1. Many of the hormones produced by the anterior pituitary are tropic hormones (tropins), hormones that stimulate other endocrine glands to secrete their hormones.

■ Communication between the hypothalamus and the posterior pituitary occurs through neurosecretory cells that span the short distance between the hypothalamus and the posterior pituitary (through the infundibulum). Hormones produced by the cell bodies of the neurosecretory cells are packaged in vesicles and transported through the axon, and stored in the axon terminals that lie in the posterior pituitary. When the neurosecretory cells are stimulated, the action potential generated triggers the release of the stored hormones from the axon terminals to a capillary network within the posterior pituitary. Two hormones, oxytocin and antidiuretic hormone (ADH), are produced and released in this way. Their functions are summarized in Table 13-1.

Table 13-1   Hormone Functions

Abbreviation	Name	Classification	Target	Action
From the Hypothalamus				
GHRH	Growth hormone RH	PP	Anterior pituitary	Stimulates release of GH
GHIH	Growth hormone IH (somatostatin)	PP	Anterior pituitary	Inhibits release of GH
TRH	Thyrotropin RH	PP	Anterior pituitary	Stimulates release of TSH and GH
GnRH	Gonadotropin RH	PP	Anterior pituitary	Stimulates release of LH and FSH
PRH	Prolactin RH	PP	Anterior pituitary	Stimulates release of PRL
PIH	Prolactin IH (dopamine)	PP	Anterior pituitary	Inhibits release of PRL
CRH	Corticotropin RH	PP	Anterior pituitary	Stimulates release of ACTH

Key to the classification abbreviations used in Table 13-1:			
H = hormone	*RH = releasing hormone*	*IH = inhibiting hormone*	*PP = polypeptide hormone*
GP = glycoprotein hormone	*P = protein*	*AA = amino acid derivative hormone*	
S = steroid hormone	*E = eicosanoid*		

(continued)

Table 13-1 *(continued)*

Abbreviation	Name	Classification	Target	Action
From the Anterior Pituitary (Tropic Hormones)				
TSH	Thyroid stimulating hormone (thyrotropin)	GP	Thyroid gland	Stimulates secretion of T_3 and T_4
ACTH	Adrenocorticotropic hormone	PP	Adrenal cortex	Stimulates secretion of glucocorticoids
FSH	Follicle stimulating hormone	GP	Ovaries and testes	Regulates oogenesis and spermatogenesis
LH	Luteinizing hormone	GP	Ovaries and testes	Causes ovulation and release of testosterone
From the Anterior Pituitary (Not Tropic Hormones)				
PRL	Prolactin	P	Mammary glands	Stimulates production of milk
hGH	human Growth Hormone (somatotropin)	P	Bones, muscles, and cells in general	Stimulates growth
From the Posterior Pituitary				
OT	Oxytocin	PP	Uterus and mammary glands	Produces uterine contractions and release of milk
ADH	Antidiuretic hormone (vasopressin)	PP	Kidneys and sweat glands	Prevents dehydration
From the Thyroid Gland				
T_4	Thyroxine	AA	Bone and general cells	Increases metabolism
T_3	Triiodothyronine	AA	Bone and general cells	Increases metabolism
CT	Calcitonin	PP	Bone and general cells	Decreases blood Ca^{2+}

Key to the classification abbreviations used in Table 13-1:

H = hormone RH = releasing hormone IH = inhibiting hormone PP = polypeptide hormone

GP = glycoprotein hormone P = protein AA = amino acid derivative hormone

S = steroid hormone E = eicosanoid

Abbreviation	Name	Classification	Target	Action
From the Parathyroid Gland				
PTH	Parathyroid hormone	PP	Bone, kidneys, and small intestine	Increases blood Ca^{+2}
From the Adrenal Medulla				
EPI	Epinephrine (adrenaline)	AA	Blood vessels, liver, and heart	Increases blood glucose levels
NE	Norepinephrine (noradrenaline)	AA	Blood vessels, liver, and heart	Increases blood glucose levels
From the Adrenal Cortex				
—	Mineralocorticoids (aldosterone)	S	Kidneys	Allows retention of Na^+ and release of K^+; increases BP
—	Glucocorticoids (such as cortisol)	S	General body cells	Increases blood glucose levels
—	Androgens	S	General body cells	Stimulates onset of puberty
From the Pancreas				
—	Glucagon (from alpha cells)	PP	Liver	Increases blood glucose levels
—	Insulin (from beta cells)	PP	Liver, muscle, and general cells of the body.	Decreases blood glucose levels

Key to the classification abbreviations used in Table 13-1:

H = hormone   *RH* = releasing hormone   *IH* = inhibiting hormone   *PP* = polypeptide hormone

GP = glycoprotein hormone   *P* = protein   *AA* = amino acid derivative hormone

S = steroid hormone   *E* = eicosanoid

(continued)

Table 13-1 *(continued)*

Abbreviation	Name	Classification	Target	Action
–	Somatostatin (from delta cells)	PP	Alpha and beta cells, and adenohypophysis	Inhibits glucagon and insulin release and inhibits release of GH
–	Pancreatic polypeptide (from F cells)	PP	Delta cells	Inhibits somatostatin and pancreatic enzyme release

From the Ovaries

–	Estrogen	S	General cells of the body	Determines secondary sex characteristics
–	Progesterone	S	Uterus	Prepares the body for pregnancy; involved in the menstrual cycle
–	Relaxin	PP	Pelvis and cervix	Prepares the birth canal and cervix dilation
–	Inhibin	P	Anterior pituitary	Inhibits release of FSH

From the Testes

–	Testosterone	S	General body cells	Determines secondary sex characteristics and plays minor role in spermatogenesis
–	Inhibin	P	Anterior pituitary	Inhibits release of FSH

From the Pineal Gland

–	Melatonin	AA	General body cells	Regulates the biological clock and circadian rhythms

Key to the classification abbreviations used in Table 13-1:

H = hormone   RH = releasing hormone   IH = inhibiting hormone   PP = polypeptide hormone

GP = glycoprotein hormone   P = protein   AA = amino acid derivative hormone

S = steroid hormone   E = eicosanoid

Abbreviation	Name	Classification	Target	Action

From the Kidneys

Abbreviation	Name	Classification	Target	Action
EPO	Erythropoietin	GP	Bone marrow	Stimulates RBC formation
—	Calcitriol (vitamin D)	S	Small intestines	Increases Ca^{+2} absorption

From the Placenta

Abbreviation	Name	Classification	Target	Action
—	Estrogen	S	Uterus	Maintains pregnancy
—	Progesterone	S	Endometrium of the uterus	Maintains pregnancy
hCG	human Chorionic Gonadotropin	GP	Corpus luteum of the ovaries	Stimulates release of estrogen and progesterone
hPL	human Placental Lactogen	P	Mammary glands	Prepares the mammary glands for lactation

From the Gastrointestinal Tract

Abbreviation	Name	Classification	Target	Action
—	Gastrin	PP	Chief (C) and Parietal (P) cells of the stomach	Stimulates release of HCl and pepsinogen, respectively
GIP	Glucose-dependent insulinotropic peptide.	PP	Beta cells of the pancreas	Stimulates the release of insulin
—	Secretin	PP	Pancreas and liver	Stimulates production of buffers from the pancreas and bile from the liver
CCK	Cholecystokinin	P	Pancreas and gallbladder	Stimulates release of enzymes from the pancreas and bile from the gallbladder
—	Serotonin	AA	Stomach	Stimulates stomach muscle contraction

Key to the classification abbreviations used in Table 13-1:

H = hormone   RH = releasing hormone   IH = inhibiting hormone   PP = polypeptide hormone
GP = glycoprotein hormone   P = protein   AA = amino acid derivative hormone
S = steroid hormone   E = eicosanoid

(continued)

Table 13-1 *(continued)*

From the Heart

ANP	Atrial natriuretic peptide	P	Kidneys	Causes loss of Na⁺ and water; decreases BP

From Adipocytes

–	Adipokines	E	General cells of the body	Is involved in anti-inflammatory responses and insulin sensitizing for increased glucose intake

From General Cells

PG	Prostaglandins	E	General cells of the body	Is involved in numerous tasks, including the inflammatory response
LT	Leukotrienes	E	General cells of the body	Is involved in numerous tasks, including the inflammatory response

Key to the classification abbreviations used in Table 13-1:

H = hormone   RH = releasing hormone   IH = inhibiting hormone   PP = polypeptide hormone

GP = glycoprotein hormone   P = protein   AA = amino acid derivative hormone

S = steroid hormone   E = eicosanoid

Endocrine Organs and Tissues

A summary of the various endocrine organs, their hormones, and their functions is given in Table 13-1. Also listed are some organs whose major function is not the secretion of hormones, but which, nonetheless, contain some specialized cells that produce hormones. These organs include the heart, the gastrointestinal tract, the placenta, the kidneys, and the skin.

In addition, all cells (except red blood cells) secrete a class of hormones called **eicosanoids.** These hormones are **paracrines,** or local hormones, that primarily affect neighboring cells. Two groups of eicosanoids, the prostaglandins (PGs) and the leukotrienes (LTs), have a wide range of varying effects that depend on the nature of the target cell. Eicosanoid activity, for example, may impact blood pressure, blood clotting, immune and inflammatory responses, reproductive processes, and the contraction of smooth muscles.

Antagonistic Hormones

Maintaining homeostasis often requires conditions to be limited to a narrow range. When conditions exceed the upper limit of homeostasis, a specific action, usually the production of a hormone, is triggered. When conditions return to normal, hormone production is discontinued. If conditions exceed the lower limit of homeostasis, a different action, usually the production of a second hormone, is triggered. Hormones that act to return body conditions to within acceptable limits from opposite extremes are called **antagonistic hormones.**

The regulation of blood glucose concentration (through negative feedback) illustrates how the endocrine system maintains homeostasis by the action of antagonistic hormones. Bundles of cells in the pancreas called pancreatic islets contain two kinds of cells, alpha cells and beta cells. These cells control blood glucose concentration by producing the antagonistic hormones insulin and glucagon:

- Beta cells secrete insulin. When the concentration of blood glucose rises (after eating, for example), beta cells secrete insulin into the blood. Insulin stimulates the liver and most other body cells to absorb glucose. Liver and muscle cells convert the glucose to glycogen (for short-term storage), and adipose cells convert the glucose to fat. In response, glucose concentration decreases in the blood, and insulin secretion discontinues (through negative feedback from declining levels of glucose).

- Alpha cells secrete glucagon. When the concentration of blood glucose drops (during exercise, for example), alpha cells secrete glucagon into the blood. Glucagon stimulates the liver to release glucose. The glucose in the liver originates from the breakdown of glycogen and the conversion of amino acids and fatty acids into glucose. When blood glucose levels return to normal, glucagon secretion discontinues (negative feedback).

Another example of antagonistic hormones occurs in the maintenance of Ca^{2+} concentration in the blood. Parathyroid hormone (PTH) from the parathyroid glands increases Ca^{2+} in the blood by increasing Ca^{2+} absorption in the intestines and reabsorption in the kidneys and stimulating Ca^{2+} release from bones. Calcitonin (CT) produces the opposite effect by inhibiting the breakdown of bone matrix and decreasing the release of calcium into the blood.

Chapter Check-Out

Q&A

1. True or False: Lipid-soluble hormones bind to a receptor protein on the receptor membrane, stimulating production of a second messenger.

2. All of the following are classifications of hormones *except*
   a. polypeptide
   b. steroid
   c. thyroxine
   d. amino acid derived

3. True or False: Parathormone inhibits the breakdown of bone, decreasing the release of calcium ions into the blood.

4. Hormones that act to return the body to within acceptable limits from extreme conditions are called _____ hormones.

Answers: 1. F, **2.** c, **3.** F, **4.** antagonist

Chapter 14

THE CARDIOVASCULAR SYSTEM

Chapter Check-In

❑ Understanding the three main functions of the cardiovascular system

❑ Detailing the pathway of blood through the chambers of the heart

❑ Describing the cardiac cycle and cardiac output

❑ Knowing the three kinds of blood vessels and which direction blood flows through them

❑ Identifying the mechanisms that help regulate blood pressure

Knowing the functions of the cardiovascular system and the parts of the body that are part of it is critical in understanding the physiology of the human body. With its complex pathways of veins, arteries, and capillaries, the cardiovascular system keeps life pumping through you. The heart, blood vessels, and blood help to transport vital nutrients throughout the body as well as remove metabolic waste. They also help to protect the body and regulate body temperature.

The Functions

The cardiovascular system consists of the heart, blood vessels, and blood. This system has three main functions:

- *Transport* of nutrients, oxygen, and hormones to cells throughout the body and removal of metabolic wastes (carbon dioxide, nitrogenous wastes).

- *Protection* of the body by white blood cells, antibodies, and complement proteins that circulate in the blood and defend the body against foreign microbes and toxins. Clotting mechanisms are also present that protect the body from blood loss after injuries.

- *Regulation* of body temperature, fluid pH, and water content of cells.

The Blood

The blood consists of cells and cell fragments, called formed elements, and water with dissolved molecules, called blood plasma (see Table 14-1).

Table 14-1   Formed Elements and Plasma of Blood

Constituent	*Characteristics/Functions*
Formed Elements (45%)	
Erythrocytes (red blood cells) (98–99%)	anucleate, contain hemoglobin; O_2 and CO_2 transport
Leukocytes (white blood cells) (0.1–0.3%):	
Neutrophils (60–70%)	granulocytes, polymorphonuclear; phagocytosis, wound healing
Eosinophils (2–4%)	granulocytes, bilobed nucleus; phagocytosis
Basophils (0.5–1%)	granulocytes, 2 to 5 lobed nucleus; release histamine
Lymphocytes (20–25%)	agranulocytes, circular nucleus, T cells, B cells; immune response, antibodies
Monocytes (3–8%)	agranulocytes, large kidney-shaped nucleus; phagocytic macrophages
Thrombocytes (platelets) (1–2%)	anucleate, megakaryocyte fragments; blood clotting

Constituent	Characteristics/Functions
Blood Plasma (55%)	
Water (90%)	
Plasma Proteins (8%):	
Albumin (54%)	maintain osmotic pressure between blood and tissues
Globulins (38%)	lipid and metal ion transporters, antibodies
Fibrinogen (7%)	clotting factor
Others (1%)	enzymes, hormones, clotting factors
Electrolytes:	Na^+, K^+, Ca^{2+}, Mg^{2+}, Cl^-, HCO_3^-, SO_4^{2-}, HPO_4^{2-}
Gases:	O_2, CO_2, N_2
Nutrients:	
Glucose, other carbohydrates	sources of energy
Amino acids	protein-building blocks
Lipids	fats, steroids, phospholipids
Cholesterol	component of plasma membranes and steroid hormones
Waste products:	
Urea	from breakdown of proteins
Creatinine	from breakdown of creatine phosphate (from muscles)
Uric acid	from breakdown of nucleic acids
Bilirubin	from breakdown of hemoglobin
Hormones:	Various

Eyrthrocytes

Erythrocytes, or red blood cells (RBCs), transport oxygen (O_2) and carbon dioxide (CO_2) in the blood. Erythrocytes contain the protein hemoglobin to which both O_2 and CO_2 attach.

Mature erythrocytes lack a nucleus and most cellular organelles, thereby maximizing the cell's volume and thus its ability to carry hemoglobin and transport O_2.

Erythrocytes are shaped like flattened donuts with a depressed center (rather than a donut hole). Their flattened shape maximizes surface area for the exchange of O_2 and CO_2 and allows flexibility that permits their passage through narrow capillaries.

Hemoglobin contains both a protein portion, called globin, and nonprotein heme groups. Globin consists of four polypeptide chains, each of which contains a heme group. The heme group is a red pigment that contains a single iron atom surrounded by a ring of nitrogen-containing carbon rings. One oxygen atom attaches to the iron of each heme group, allowing a single hemoglobin molecule to carry four oxygen atoms. Each erythrocyte contains about 250 million hemoglobin molecules.

Oxyhemoglobin (HbO_2) forms in the lungs when erythrocytes are exposed to oxygen as they pass through the lungs. Deoxyhemoglobin (Hb) forms when oxygen detaches from the iron and diffuses into surrounding tissues.

Carbaminohemoglobin ($HbCO_2$) forms when CO_2 attaches to amino acids of the globin part of the hemoglobin molecule. About 20–25 percent of the CO_2 transported from tissues to lungs is in this form.

Carbonic anhydrase, an enzyme in erythrocytes, converts CO_2 and H_2O in the cells to H^+ and $HCO_3{}^-$. About 70 percent of the CO_2 collected from tissues travels in the erythrocytes as $HCO_3{}^-$. About 10 percent of the carbon dioxide stays in the plasma and is transported in the circulatory system as the bicarbonate ion.

Because they lack cellular organelles and thus the physiology to maintain themselves, erythrocytes survive for only about 120 days. Degenerated erythrocytes are broken down in the spleen and liver by macrophages (phagocytic white blood cells) as follows:

1. The globin and heme parts of the hemoglobin are separated. The globin is reduced to amino acids, which are returned to the blood plasma.

2. Iron is removed from the heme group and bound to the proteins ferritin and hemosiderin, which store the iron for later use (because unbound iron is toxic). Iron is also attached to transferrin, which enters the bloodstream. Transferrin may be picked up by muscles or liver cells, where it may be stored as ferritin or hemosiderin or picked up by bone marrow, where the iron is used to produce new erythrocytes.

3. The remainder of the heme group is broken down into **bilirubin** (a yellow-orange pigment), which enters the bloodstream and is picked up by the liver. Liver cells incorporate bilirubin into bile, which enters the small intestine during the digestion of fats. Bilirubin is then converted into **urobilinogen** by intestinal bacteria. Finally, most urobilinogen is converted to the brown pigment stercobilin, which is eliminated with the feces (and which gives feces its brown color). A small amount of urobilinogen is absorbed into the blood, converted to the yellow pigment urobilin, picked up by the kidneys, and eliminated with the urine (contributing to the yellow color of urine).

Leukocytes

Leukocytes, or white blood cells (WBCs), protect the body from foreign microbes and toxins. Although all leukocytes can be found in the blood stream, some permanently leave the bloodstream to enter tissues where they encounter microbes or toxins, while other kinds of leukocytes readily move in and out of the bloodstream. Leukocytes are classified into two groups, granulocytes and agranulocytes, based on the presence or absence of granules in the cytoplasm and the shape of the nucleus. Leukocytes have just one nucleus, but some leukocytes have a multilobed nucleus, making them look like they have several nuclei.

■ **Granulocytes** contain numerous granules in the cytoplasm and have a nucleus that is irregularly shaped with lobes. Each of the three types of granulocytes is named after the type of stain that its granules absorb:

■ *Neutrophils,* the most numerous of granulocytes, have an S- or C-shaped nucleus with three to six lobes. Their granules, which are small and inconspicuous, poorly absorb both basic and acidic stains (neutral pH preference), producing a pale, lilac color. Because the shape of the nucleus is so variable, neutrophils are referred to as polymorphonuclear leukocytes (PMNs), or polys. Young neutrophils that are shaped like rods are called band neutrophils. Older neutrophils, with a segmented nucleus, are called segs. Neutrophils are the first leukocytes to arrive at a site of infection, responding (by **chemotaxis**) to chemicals released by damaged cells. The neutrophils, by **phagocytosis,** actively engulf bacteria, which are then destroyed by the various antibiotic proteins (such as defensins and lysozymes) contained within the granules. The neutrophils, usually destroyed in the process, contribute, together with other dead tissue, to the formation of pus.

■ *Eosinophils* have a bilobed nucleus (two lobes connected by a narrow strand of chromatin). Their granules, which stain red with acid (eosin) dyes, contain digestive enzymes. Their granules are therefore considered to be lysosomes. Eosinophils actively phagocytize complexes formed by the action of antibodies on antigens (foreign substances). Numbers of eosinophils increase during parasitic infection and allergic reactions.

■ *Basophils* have a U- or S-shaped nucleus with two to five lobes connected by a narrow strand of chromatin. Their granules, which stain blue-purple with basic dyes, contain histamine, serotonin, and heparin. Basophils release histamine in response to tissue damage and to pathogen invasion (as part of the inflammatory response). Basophils resemble mast cells, which are similar in appearance and function to basophils, but found only in connective tissues. Many times, there are so many granules in a basophil that you cannot see the nucleus.

■ **Agranulocytes,** the second group of leukocytes, do not have visible granules in the cytoplasm and the nucleus is not lobed. There are two types of these leukocytes:

■ *Lymphocytes,* often classified as small, medium, and large, have a roughly round nucleus surrounded by a small amount of blue-staining cytoplasm. Lymphocytes are the only leukocytes that return to the bloodstream, circulating among the bloodstream, tissue fluids, tissues, and lymph fluid. There are two major groups of lymphocytes, which vary based upon their role in an immune response. T lymphocytes (**T cells**), which mature in the thymus gland, and upon exposure to thymosin attack aberrant cells (such as tumor cells, organ transplant cells, or cells infected by viruses). B lymphocytes (B cells), which mature in the bone marrow, respond to circulating antigens (such as toxins, viruses, or bacteria) by dividing to produce plasma cells, which in turn produce antibodies.

■ *Monocytes* have a large, kidney-shaped nucleus surrounded by ample blue-gray-staining cytoplasm. When monocytes leave the bloodstream and move into tissues, they enlarge and become macrophages, which engulf microbes and cellular debris.

Platelets

Platelets (**thrombocytes**) are fragments of huge cells called megakaryocytes. Megakaryocytes fragment as they pass from the bone marrow into the bloodstream. Platelets lack a nucleus and consist of cytoplasm (with few organelles) surrounded by a plasma membrane. Platelets adhere to damaged blood vessel walls and release enzymes that activate **hemostasis,** the stoppage of bleeding.

Plasma

Plasma is the straw-colored, liquid portion of the blood. It consists of the following:

- Water (90 percent).

- Proteins (8 percent). Albumin, the most common protein, is produced by the liver and serves to preserve osmotic pressure between blood and tissues. Other proteins include alpha and beta globulins (proteins that transport lipids and metal ions), gamma globulins (antibodies), fibrinogen and prothrombin (clotting proteins), and hormones.

- Waste products (urea, uric acid, creatinine, bilirubin, and others).

- Nutrients (absorbed from the digestive tract).

- Electrolytes (various ions such as sodium, calcium, chloride, and bicarbonate).

- Respiratory gases (O_2 and CO_2).

Serum is the liquid material remaining after blood-clotting proteins have been removed from plasma as a result of clotting.

Blood Formation

Hemopoiesis (**hematopoiesis**) is the process that produces the formed elements of the blood. Hemopoiesis takes place in the red bone marrow found in the epiphyses of long bones (for example, the humerus and femur), flat bones (ribs and cranial bones), vertebrae, and the pelvis. Within the red bone marrow, hemopoietic stem cells (**hemocytoblasts**) divide to produce various "blast" cells. Each of these cells matures and becomes a particular formed element.

Erythropoiesis

Erythropoiesis, the process of making erythrocytes, begins with the formation of proerythroblasts from hemopoietic stem cells. Over three to five days, several stages of development follow as ribosomes proliferate and hemoglobin is synthesized. Finally, the nucleus is ejected, producing the depression in the center of the cell. Young erythrocytes, called reticulocytes, still containing some ribosomes and endoplasmic reticulum, pass into the bloodstream and develop into mature erythrocytes after another one or two days.

Erythropoietin

Erythropoietin (EPO), a hormone produced mostly by the kidneys, stimulates bone marrow to produce erythrocytes (stimulates **erythropoiesis**). When inadequate amounts of oxygen are delivered to body cells, a condition called hypoxia, the kidneys increase EPO secretion, which in turn stimulates an increase in erythrocyte production.

The average production rate of erythrocytes in healthy individuals is two million cells per second. Normal production requires adequate amounts of iron, vitamin B_{12}, and folic acid. Vitamin B_{12} and folic acid are necessary for the proper development of DNA in the erythroblasts. This DNA is responsible for the organization of the heme molecule of which iron will become a component. Proper DNA development is also necessary for erythroblast reproduction. A lack of either vitamin B_{12} or folic acid can result in pernicious anemia.

Leukopoiesis

Leukopoiesis, the process of making leukocytes, is stimulated by various colony-stimulating factors (CSFs), which are hormones produced by mature white blood cells. The development of each kind of white blood cell begins with the division of the hemopoietic stem cells into one of the following "blast" cells:

- *Myeloblasts* divide to form eosinophilic, neutrophilic, or basophilic myelocytes, which lead to the development of the three kinds of granulocytes.

- *Monoblasts* lead to the development of monocytes.

- *Lymphoblasts* lead to the development of lymphocytes.

Thrombopoiesis

Thrombopoiesis, the process of making platelets, begins with the formation of **megakaryoblasts** from hemopoietic stem cells. The megakaryoblasts divide without cytokinesis to become megakaryocytes, huge cells with a large, multilobed nucleus. The megakaryocytes then fragment into segments as the plasma membrane infolds into the cytoplasm.

Hemostasis

Hemostasis, the stoppage of bleeding, is accomplished through three steps:

1. A vascular spasm, a constriction of the damaged blood vessel, occurs at the site of injury. **Vasoconstriction** is initiated by the smooth muscle of the blood vessel in response to the injury and by nerve signals from pain receptors.

2. A platelet plug, consisting of a mass of linked platelets, fills the hole in the damaged blood vessel. Platelet plug formation follows these steps:

   a. *Platelet adhesion.* Platelets adhere to the exposed collagen fibers in the damaged blood vessel wall.

   b. *Platelet release.* Platelets release ADP (adenosine diphosphate, which attracts other platelets to the injury), serotonin (which stimulates vasoconstriction), and thromboxane A_2 (which attracts platelets and stimulates vasoconstriction, and keeps the platelets "sticky" so they continue to adhere to the injured site). Cellular extensions from the platelets interconnect and form a loose mesh. Aspirin inhibits the formation of clots because aspirin prevents the formation of thromboxane A_2.

   c. *Platelet aggregation.* Additional platelets arrive at the site of the injury in response to the released ADP and expand the accumulation of platelets.

3. Coagulation (blood clotting) is a complex series of reactions that transform liquid blood into a gel (clot), providing a secure patch to the injured blood vessel. Thirteen coagulation factors (numbered I through XIII in order of their discovery) are involved. Most of these factors are proteins released into the blood by the liver. Factor IV is Ca^{2+}. Vitamin K is required for the synthesis of some of these factors. The coagulation process can be described in three major steps:

a. *Formation of factor X and prothrombinase.* Prothrombinase (prothrombin activator) can form either intrinsically (inside the blood vessels) or extrinsically (outside the blood vessels). In the intrinsic pathway, the collagen of the damaged blood vessel initiates a cascade of reactions that activate factor X. In the extrinsic pathway, damaged tissues release thromboplastin (tissue factor, TF), which initiates a shorter and more rapid sequence of reactions to activate factor X. In both pathways, activated factor X combines with factor V (with Ca^{2+} present) to form prothrombinase.

b. *Prothrombin is converted to thrombin.* In this common pathway that follows both the intrinsic and extrinsic pathways, prothrombinase (with Ca^{2+}) converts prothrombin to thrombin.

c. *Fibrinogen is converted to fibrin.* The common pathway continues as thrombin (with Ca^{2+}) converts fibrinogen to fibrin. Fibrin forms long strands that bind the platelets together to form a dense web. Thrombin also activates factor XIII, which helps fibrin strands cohere to one another. The result is a clot. See Figure 14-1.

Figure 14-1 The Clotting Process

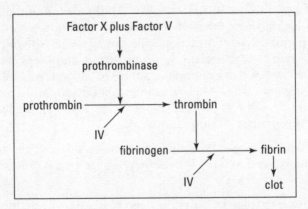

Following its formation, a clot is further strengthened by a process called clot retraction. Platelets in the clot contract, pulling on the fibrin strands to which they are attached. The result is a more tightly sealed patch.

Fibrinolysis is the breakdown of the clot as the damaged blood vessel is repaired. During the formation of a clot, the plasma protein plasminogen is incorporated into the clot. The healthy endothelial tissue that replaces the damaged areas of the blood vessel secretes tissue plasminogen activator (t-PA), which converts plasminogen into its active form, plasmin (fibrinolysin). Plasmin in turn breaks down fibrin and leads to the dissolution of the clot.

Blood Groups

Various glycoproteins and lipoproteins are embedded in the surfaces of red blood cells. These proteins are inherited, and their structures may vary from one individual to another. If during a transfusion an individual receives blood containing RBCs with proteins that the individual does not carry, these proteins may be recognized as foreign antigens by the immune system. If so, antibodies are produced that bind to the antigens and cause **agglutination** (clumping) and subsequent destruction of the foreign RBCs.

There are over 30 common groups of RBC proteins, referred to as antigens, isoantigens, or **agglutinogens** (which is the preferred term). Generally, each group is controlled by a single gene, and for each gene, two alleles, or forms, of the gene are inherited (one allele from each parent). Each blood group gene may have two or more different alleles in the population. Although not all blood group proteins stimulate the immune response, two important ones do:

- ABO blood group. The gene responsible for this group has three alleles. One allele produces an "A" agglutinogen, a second produces a "B" agglutinogen, and a third produces no agglutinogens ("O"). Because individuals inherit two alleles, individuals may be of the A blood type, inheriting two A alleles (AA) or an A and an O allele (AO); the B blood type (BB or BO); the AB blood type (AB); or the O blood type (OO). The immune response is activated when an individual receives a transfusion with blood carrying nonself agglutinogens. For example, the immune system would respond if a person with A blood type (either AA or AO) receives blood of the B or AB blood type, but not of the O type. (The O type does not carry any foreign agglutinogens.)

■ **Rh blood group.** This is a complex group defined by agglutinogens produced by three different genes. Each gene has two (or rarely, three) alleles. Because of the close linkage of the genes (they are positioned close to one another on the same chromosome), the expression of the group can be evaluated as if it were a single gene with two alleles, an Rh⁺ allele (producing the Rh agglutinogen) and an Rh⁻ allele (producing no Rh agglutinogen). Thus, individuals are either Rh⁺ if they inherit one or two Rh⁺ alleles or Rh⁻ if they inherit two Rh⁻ alleles. The Rh factor is typically called the D agglutinogen. It was originally found on the red blood cells of Rhesus monkeys (hence the "Rh" factor).

Circulatory Pathways

Blood is confined to a closed system of blood vessels and to the four chambers of the heart (essentially dilated vessels). Blood travels away from the heart through arteries, which branch into smaller vessels, the arterioles. Arterioles branch further into the smallest vessels, the capillaries. Gas, nutrient, and waste exchange occurs across the capillary walls. The blood returns to the heart as capillaries merge to form venules, which further merge to form large veins, which connect to the heart. Blood circulates through the following two separate circuits:

■ In the pulmonary circulation, deoxygenated blood travels from the right side of the heart to each of the two lungs. Within the lungs, O_2 enters and CO_2 leaves the capillaries by diffusion. Oxygenated blood returns from the lungs to the left side of the heart.

■ In the systemic circulation, oxygenated blood travels from the left side of the heart to the various areas of the body. Gas, nutrient, and waste exchange occurs across the capillary walls into the interstitial fluids outside the capillaries and then into the surrounding cells. The deoxygenated blood returns to the right side of the heart.

The Heart

The heart is located in the mediastinum, the cavity between the lungs. The heart is tilted so that its pointed end, the apex, points downward toward the left hip, while the broad end, the base, faces upward toward the right shoulder. The heart is surrounded by the pericardium, a sac characterized by the following two layers:

■ The outer fibrous pericardium anchors the heart to the surrounding structures.

■ The inner serous pericardium consists of an outer parietal layer and an inner visceral layer. A thick layer of serous fluid, the pericardial fluid, lies between these two layers to provide a slippery surface for the movements of the heart.

The wall of the heart consists of three layers:

■ The *epicardium* is the visceral layer of the serous pericardium.

■ The *myocardium* is the muscular part of the heart that consists of contracting cardiac muscle and noncontracting Purkinje fibers that conduct nerve impulses. Cardiac cells (cardiomyocytes) are in this layer.

■ The *endocardium* is the thin, smooth, endothelial, inner lining of the heart, which is continuous with the inner lining of the blood vessels.

As blood travels through the heart, it enters a total of four chambers and passes through four valves. The two upper chambers, the right and left atria, are separated longitudinally by the interatrial septum. The two lower chambers, the right and left ventricles, are the pumping machines of the heart and are separated longitudinally by the interventricular septum. A valve follows each chamber and prevents the blood from flowing backward into the chamber from which the blood originated.

Two prominent grooves are visible on the surface of the heart:

■ The coronary sulcus (atrioventricular groove) marks the junction of the atria and ventricles.

■ The anterior interventricular sulcus and posterior interventricular sulcus mark the junction of the ventricles on the anterior and posterior sides of the heart, respectively.

The pathway of blood through the chambers and valves of the heart is described as follows (see Figure 14-2):

Figure 14-2 The pathway of blood through the chambers and valves of the heart.

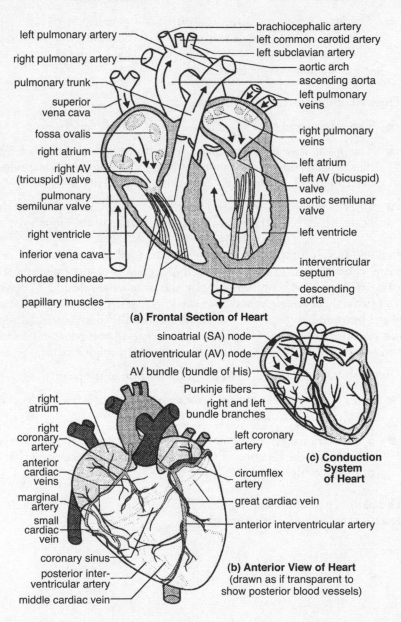

left pulmonary artery

right pulmonary artery

pulmonary trunk

superior vena cava

fossa ovalis

right atrium

right AV (tricuspid) valve

pulmonary semilunar valve

right ventricle

inferior vena cava

chordae tendineae

papillary muscles

brachiocephalic artery
left common carotid artery
left subclavian artery
aortic arch
ascending aorta
left pulmonary veins
right pulmonary veins
left atrium
left AV (bicuspid) valve
aortic semilunar valve
left ventricle
interventricular septum
descending aorta

(a) Frontal Section of Heart

sinoatrial (SA) node
atrioventricular (AV) node
AV bundle (bundle of His)
Purkinje fibers
right and left bundle branches

(c) Conduction System of Heart

right atrium

right coronary artery

anterior cardiac veins

marginal artery

small cardiac vein

coronary sinus

posterior interventricular artery

middle cardiac vein

left coronary artery

circumflex artery

great cardiac vein

anterior interventricular artery

(b) Anterior View of Heart
(drawn as if transparent to show posterior blood vessels)

- The right atrium, located in the upper right side of the heart, and a small appendage, the right auricle, act as a temporary storage chamber so that blood will be readily available for the right ventricle. Deoxygenated blood from the systemic circulation enters the right atrium through three veins: the superior vena cava, the inferior vena cava, and the coronary sinus. During the interval when the ventricles are not contracting, blood passes down through the right atrioventricular (AV) valve into the next chamber, the right ventricle. The AV valve is also called the tricuspid valve because it consists of three flexible cusps (flaps).

- The right ventricle is the pumping chamber for the pulmonary circulation. The ventricle, with walls thicker and more muscular than those of the atrium, contracts and pumps deoxygenated blood through the three-cusped pulmonary semilunar valve and into a large artery, the pulmonary trunk. The pulmonary trunk immediately divides into two pulmonary arteries, which lead to the left and right lungs, respectively. The following events occur in the right ventricle:

  - When the right ventricle contracts, the right AV valve closes and prevents blood from moving back into the right atrium. Small tendonlike cords, the chordae tendineae, are attached to papillary muscles at the opposite, bottom side of the ventricle. These cords limit the extent to which the AV valve can be forced closed, preventing it from being pushed through and into the atrium.

  - When the right ventricle relaxes, there is less pressure in the right ventricle and more pressure in the pulmonary trunk. This high pressure in the pulmonary trunk causes the valve to close, thereby preventing the backflow of blood and the return of blood to the right ventricle.

- The left atrium and its auricle appendage receive oxygenated blood from the lungs through four pulmonary veins (two from each lung). The left atrium, like the right atrium, is a holding chamber for blood in readiness for its flow into the left ventricle. When the ventricles relax, blood leaves the left atrium and passes through the left AV valve into the left ventricle. The left AV valve is also called the mitral or bicuspid valve, the only heart valve with two cusps.

- The left ventricle is the pumping chamber for the systemic circulation. Because a greater blood pressure is required to pump blood through the much more extensive systemic circulation than through the pulmonary circulation, the left ventricle is larger and its walls are thicker than those of the right ventricle. When the left ventricle contracts, it pumps oxygenated blood through the aortic semilunar valve, into a large artery, the aorta, and throughout the body. The following events occur in the left ventricle, simultaneously and analogously with those of the right ventricle:

  1. When the left ventricle contracts, the left AV valve closes and prevents blood from moving back into the right atrium. As in the right AV valve, the chordae tendineae prevent overextension of the left AV valve.

  2. When the left ventricle relaxes, this results in less pressure in the left ventricle and higher pressure in the aorta. This high pressure causes the aortic semilunar valve to close, thus preventing the return of blood to the left ventricle.

Two additional passageways are present in the fetal heart:

- The foramen ovale is an opening across the interatrial septum. It allows blood to bypass the right ventricle and the pulmonary circuit while the nonfunctional fetal lungs are still developing. The opening, which closes at birth, leaves a shallow depression called the fossa ovalis in the adult heart.

- The ductus arteriosus is a connection between the pulmonary trunk and the aorta. Blood that enters the right ventricle is pumped out through the pulmonary trunk. Although some blood enters the pulmonary arteries (to provide oxygen and nutrients to the fetal lungs), most of the blood moves directly into the aorta through the ductus arteriosus.

The coronary circulation consists of blood vessels that supply oxygen and nutrients to the tissues of the heart. Blood entering the chambers of the heart cannot provide this service because the endocardium is too thick for effective diffusion (and only the left side of the heart contains oxygenated blood). Instead, the following two arteries that arise from the aorta and encircle the heart in the artioventricular groove provide this function:

- The left coronary artery has the following two branches: the anterior interventricular artery (left anterior descending, or LAD, artery) and the circumflex artery.

- The right coronary artery has the following two branches: the posterior interventricular artery and the marginal artery.

Blood from the coronary circulation returns to the right atrium by way of an enlarged blood vessel, the coronary sinus. Three veins, the great cardiac vein, the middle cardiac vein, and the small cardiac vein, feed the coronary sinus.

Cardiac Conduction

Unlike skeletal muscle fibers (cells), which are independent of one another, cardiac muscle fibers (contractile muscle fibers) are linked by intercalated discs, areas where the plasma membranes intermesh. Within the intercalated discs, the adjacent cells are structurally connected by desmosomes, tight seals that weld the plasma membranes together, and electrically connected by gap junctions, ionic channels that allow the transmission of a depolarization event. As a result, the entire myocardium functions as a single unit with a single contraction of the atria followed by a single contraction of the ventricles.

Action potentials (electrical impulses) in the heart originate in specialized cardiac muscle cells, called **autorhythmic cells.** These cells are self-excitable, able to generate an action potential without external stimulation by nerve cells. The autorhythmic cells serve as a pacemaker to initiate the cardiac cycle (pumping cycle of the heart) and provide a conduction system to coordinate the contraction of muscle cells throughout the heart. The autorhythmic cells are concentrated in the following areas (refer to Figure 14-2):

- The sinoatrial (SA) node, located in the upper wall of the right atrium, initiates the cardiac cycle by generating an action potential that spreads through both atria through the gap junctions of the cardiac muscle fibers.

- The atrioventricular (AV) node, located near the lower region of the interatrial septum, receives the action potential generated by the SA node. A slight delay of the electrical transmission occurs here, allowing the atria to fully contract before the action potential is passed on to the ventricles.

■ The atrioventricular (AV) bundle (bundle of His) receives the action potential from the AV node and transmits the impulse to the ventricles by way of the right and left bundle branches. Except for the AV bundle, which provides the only electrical connection, the atria are electrically insulated from the ventricles.

■ The Purkinje fibers are large-diameter fibers that conduct the action potential from the interventricular septum, down to the apex, and then upward through the ventricles.

Cardiac Muscle Contraction

The sarcolemma (plasma membrane) of an unstimulated muscle cell is polarized—that is, the inside of the sarcolemma is negatively charged with respect to the outside. The unstimulated state of the muscle cell, called the resting potential, is created by the presence of large, negatively charged proteins and nucleic acids inside the cell. A balance between K^+ inside the cell and Na^+ outside the cell contributes to the polarization. During an action potential, the balance of Na^+ and K^+ is upset so that the cell becomes depolarized. The series of events that occurs during and following an action potential in contractile muscle fibers of the heart is similar to that in skeletal muscle. Here is a description of these events:

1. Rapid depolarization occurs when fast-opening Na^+ channels in the sarcolemma open and allow an influx of Na^+ ions into the cardiac muscle cell. The Na^+ channels rapidly close.

2. A plateau phase occurs during which Ca^{2+} enters the cytosol of the muscle cell. Ca^{2+} enters from the sarcoplasmic reticulum (endoplasmic reticulum) within the cell and also from outside the cell through slow-opening Ca^{2+} channels in the sarcolemma. Within the cell, Ca^{2+} binds to troponin, which in turn triggers the cross-bridge binding that leads to the sliding of actin filaments past myosin filaments. The sliding of the filaments produces cell contraction. At the same time that the Ca^{2+} channels open, K^+ channels, which normally leak small amounts of K^+ out of the cell, become more impermeable to K^+ leakage. The combined effects of the prolonged release of Ca^{2+} and the restricted leakage of K^+ lead to an extended depolarization that appears as a plateau when membrane potential is plotted against time.

3. Repolarization occurs as K^+ channels open and K^+ diffuses out of the cell. At the same time, Ca^{2+} channels close. These events restore the membrane to its original polarization, except that the positions of K^+ and Na^+ on each side of the sarcolemma are reversed.

4. A refractory period follows, during which concentration of K^+ and Na^+ are actively restored to their appropriate sides of the sarcolemma by Na^+/K^+ pumps. The muscle cell cannot contract again until Na^+ and K^+ are restored to their resting potential states. The refractory period of cardiac muscle is dramatically longer than that of skeletal muscle. This prevents tetanus from occurring and ensures that each contraction is followed by enough time to allow the heart chamber to refill with blood before the next contraction.

Electrocardiogram

Electrical currents generated by the heart during the cardiac cycle can be detected on the surface of the body by the electrodes of an electrocardiograph. A recording of these currents, called an electrocardiogram (ECG or EKG), represents a sum of all the concurrent action potentials produced by the heart as detected by the 12 electrodes of the electrocardiograph. A single cardiac cycle produces a distinctive wave pattern, where peaks and valleys are indicated by the letters P, Q, R, S, and T (see Figure 14-3). An interpretation of the major characteristics of the ECG follows:

■ The P wave is a small wave that represents the depolarization of the atria. During this wave, the muscles of the atria are contracting.

■ The QRS complex is a rapid down-up-down movement. The upward movement produces a tall peak, indicated by R. The QRS complex represents the depolarization of the ventricles.

■ The T wave represents the repolarization of the ventricles. Electrical activity generated by the repolarization of the atria is concealed by the QRS complex.

Figure 14-3 Different periods of the cardiac cycle: (a) electrocardiogram, (b) heart sounds, (c) valves, (d) pressure, and (e) volume of left ventricle.

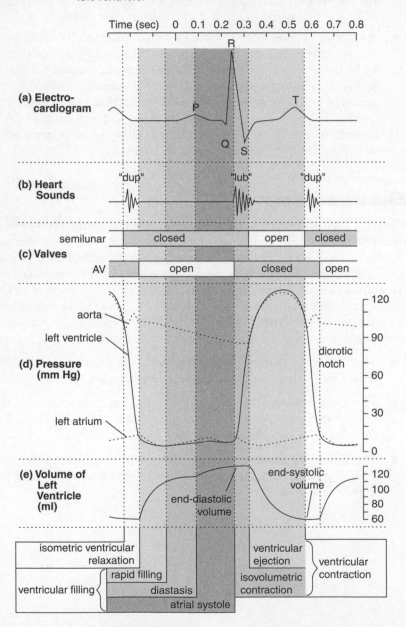

The Cardiac Cycle

The cardiac cycle describes all the activities of the heart through one complete heartbeat—that is, through one contraction and relaxation of both the atria and ventricles. A contraction event (of either the atria or ventricles) is referred to as **systole,** and a relaxation event is referred to as **diastole.** The cardiac cycle includes a description of the systolic and diastolic activities of the atria and ventricles, the blood volume and pressure changes within the heart, and the action of the heart valves. A description of each period of the cardiac cycle follows (also refer to Figure 14-3):

■ The isovolumetric ventricular relaxation is the period during which the ventricles are relaxed and both AV and semilunar valves are still closed. The volume of the ventricles remains unchanged (isovolumetric) during this period.

■ Ventricular filling begins as the AV valves open and blood fills the ventricles. The ventricles remain in diastole during this period. The filling of the ventricles can be described as three successive events:

■ Rapid ventricular filling occurs as blood flows into the empty and relaxed ventricles. Volume of the ventricles increases rapidly.

■ Diastasis is a slower filling event than that of the preceding because most of the volume of the ventricle is already occupied by blood.

■ Atrial systole (and the P wave of the ECG) occurs and forces the remaining blood from the atria into the ventricles. The blood volume at the end of this interval is called the **end-diastolic volume (EDV).**

■ Ventricular contraction (ventricular systole) begins as the action potential from the AV node enters the ventricles, the ventricles depolarize, and the QRS complex is observed on the ECG. The following intervals during this phase are observed:

■ Isovolumetric contraction occurs when the AV valves are forced shut. During this brief period, while the semilunar valves are still closed, the volume of the ventricles remains unchanged.

■ Ventricular ejection occurs as the continuing contraction of the ventricles increases the pressure in the ventricles and forces the semilunar valves open. At this point, blood is forced out of the ventricles. This interval ends when the ventricles begin to relax, and the semilunar valves close. The closing of the semilunar valves causes a small increase in blood pressure visible as the dicrotic notch on a plot of blood pressure against time. The amount of blood remaining in the ventricles at this time is called the **end-systolic volume (ESV).**

The heart sounds associated with the beating of the heart can be heard by auscultating (listening to) the thorax with a stethoscope. The two major heart sounds, described as "lub-dub," originate from blood turbulence generated by the closing of the AV valves and the semilunar valves, respectively (refer to Figure 14-3). Abnormal heart sounds called murmurs are usually caused by improperly functioning valves.

Cardiac Output

The following variables are measures of the capacity of the heart:

■ **Stroke volume (SV)** is the volume of blood ejected by each ventricle during a single contraction.

■ **Heart rate (HR)** is the number of heartbeats per minute.

■ **Cardiac output (CO)** is the volume of blood pumped out of the right or left ventricle per minute. $CO = SV \times HR$.

Cardiac output varies widely with the health of the individual and the state of activity at the time of measurement. Cardiac output in exercising athletes may exceed their resting cardiac output seven times. The ratio between the maximum and resting cardiac output of an individual is the cardiac reserve. Note that cardiac output changes when either stroke volume or heart rate changes.

Stroke volume is regulated by the following three factors:

■ Preload is the degree to which cardiac muscle cells are stretched by the blood entering the heart chambers. According to the Frank-Starling law of the heart, the more the chamber is stretched, the greater the force of its contraction. Because the end-diastolic volume (EDV) is a measure of how much blood enters the ventricles, the EDV is an indicator of ventricle preload.

- Contractility is the degree to which cardiac muscle cells contract as a result of extrinsic influences. Positive inotropic factors, such as certain hormones (epinephrine or thyroxin), drugs (digitalis), or elevated levels of Ca^{2+}, increase contractility, while negative inotropic factors, such as certain drugs (calcium channel blockers) or elevated levels of K^+, decrease contractility.

- **Afterload** is a measure of the pressure that must be generated by the ventricles to force the semilunar valves open. The greater the afterload, the smaller the stroke volume. **Arteriosclerosis** (narrowing of the arteries) and high blood pressure increase afterload and reduce stroke volume.

Heart rate is regulated by the following three factors:

- The autonomic nervous system may influence heart rate when the sympathetic nervous system stimulates cardiac muscle contractions or when the parasympathetic system inhibits cardiac muscle contractions.

- Chemicals such as hormones and ions can influence heart rate. Epinephrine, secreted by the adrenal medulla, and thyroxin, secreted by the thyroid gland, increase heart rate. Abnormal blood concentrations of Na^+, K^+, and Ca^{2+} interfere with muscle contraction.

- Other factors such as age, gender, body temperature, and physical fitness may influence heart rate.

Blood Vessels

The central opening of a blood vessel, the lumen, is surrounded by a wall consisting of three layers:

- The tunica intima is the inner layer facing the blood. It is composed of an innermost layer of endothelium (simple squamous epithelium) surrounded by variable amounts of connective tissues.

- The tunica media, the middle layer, is composed of smooth muscle with variable amounts of elastic fibers.

- The tunica adventitia, the outer layer, is composed of connective tissue.

The cardiovascular system consists of three kinds of blood vessels that form a closed system of passageways:

- Arteries carry blood away from the heart. The three kinds of arteries are categorized by size and function:

  - Elastic arteries (conducting arteries) are the largest arteries and include the aorta and other nearby branches. The tunica media of elastic arteries contains a large amount of elastic connective tissue, which enables the artery to expand as blood enters the lumen from the contracting heart. During relaxation of the heart, the elastic wall of the artery recoils to its original position, forcing blood forward and smoothing the jerky discharge of blood from the heart.

  - Muscular arteries (conducting arteries) branch from elastic arteries and distribute blood to the various body regions. Abundant smooth muscle in the thick tunica media allows these arteries to regulate blood flow by **vasoconstriction** (narrowing of the lumen) or **vasodilation** (widening of the lumen). Most named arteries of the body are muscular arteries.

  - Arterioles are small, nearly microscopic blood vessels that branch from muscular arteries. Most arterioles have all three tunics present in their walls, with considerable smooth muscle in the tunica media. The smallest arterioles consist of endothelium surrounded by a single layer of smooth muscle. Arterioles regulate the flow of blood into capillaries by vasoconstriction and vasodilation.

- Capillaries are microscopic blood vessels with extremely thin walls. Only the tunica intima is present in these walls, and some walls consist exclusively of a single layer of endothelium. Capillaries penetrate most body tissues with dense interweaving networks called capillary beds. The thin walls of capillaries allow the diffusion of oxygen and nutrients out of the capillaries, while allowing carbon dioxide and wastes into the capillaries.

  Below is a list of the different types of capillaries:

  - Metarterioles (precapillaries) are the blood vessels between arterioles and venules. Although metarterioles pass through capillary beds with capillaries, they are not true capillaries because metarterioles, like arterioles, have smooth muscle present in the tunica media. The smooth muscle of a metarteriole allows it to act as a shunt to regulate blood flow into the true capillaries that branch from it. The thoroughfare channel, the tail end of the metarteriole that connects to the venule, lacks smooth muscle.

- True capillaries form the bulk of the capillary bed. They branch away from a metarteriole at its arteriole end and return to merge with the metarteriole at its venule end (thoroughfare channel). Some true capillaries connect directly from an arteriole to a metarteriole or venule. Although the walls of true capillaries lack muscle fibers, they possess a ring of smooth muscle called a pre-capillary sphincter where they emerge from the metarteriole. The precapillary sphincter regulates blood flow through the capillary. There are three types of true capillaries:

- Continuous capillaries have continuous, unbroken walls consisting of cells that are connected by tight junctions. Most capillaries are of this type.

- Fenestrated capillaries have continuous walls between endothelial cells, but the cells have numerous pores (fenestrations) that increase their permeability. These capillaries are found in the kidneys, lining the small intestine, and in other areas where a high transfer rate of substances into or out of the capillary is required.

- Sinusoidal capillaries (sinusoids) have large gaps between endothelial cells that permit the passage of blood cells. These capillaries are found in the bone marrow, spleen, and liver.

- **Veins** carry blood toward the heart. The three kinds of veins are listed here in the sequence they occur regarding the flow of blood back to the heart:

- Postcapillary venules, the smallest veins, form when capillaries merge as they exit a capillary bed. Much like capillaries, they are very porous, but with scattered smooth muscle fibers in the tunica media.

- Venules form when postcapillary venules join. Although the walls of larger venules contain all three layers, they are still porous enough to allow white blood cells to pass.

- Veins have walls with all three layers, but the tunica intima and tunica media are much thinner than in similarly sized arteries. Few elastic or muscle fibers are present. The wall consists primarily of a well-developed tunica adventitia. Many veins, especially those in the limbs, have valves, formed from folds of the tunica intima, that prevent the backflow of blood. If these valves fail to close properly, varicose veins may occur.

Many regions of the body receive blood supplies from two or more arteries. The points where these arteries merge are called arterial **anastomoses.** Arterial anastomoses allow tissues to receive blood even after one of the arteries supplying blood has been blocked.

Blood Pressure

Hydrostatic pressure created by the heart forces blood to move through the arteries. Systolic blood pressure, the pressure measured during contraction of the ventricles, averages about 110 mm Hg in arteries of the systemic circulation (for healthy, young adults). The diastolic blood pressure, measured during ventricle relaxation, is about 75 mm Hg in these arteries. As blood travels through the arterial system, resistance from the walls of the blood vessels reduces the pressure and velocity of the blood (see Figure 14-4). Blood pressure drops sharply in the arterioles and falls to between 40 and 20 mm Hg in the capillaries. Blood pressure descends further in the venules and approaches zero in the veins.

Figure 14-4   As blood travels through the arterial system, resistance from the walls of the blood vessels reduces the pressure and velocity of the blood.

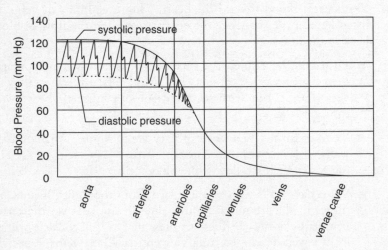

Because blood pressure is so low in venules and veins, two mechanisms assist the return of blood to the heart (venous return):

- The muscular pump arises from contractions of skeletal muscles surrounding the veins. The contractions squeeze the veins, forcing the blood to move forward, the only direction it can move when valves in the veins close to prevent backflow.

- The respiratory pump is created by the expansion and contraction of the lungs during breathing. During inspiration (inhaling), pressure in the abdominal region increases while pressure in the thoracic cavity decreases. These pressures act upon the veins passing through these regions. As a result, blood flows toward the heart as it moves from regions of higher pressure (the abdomen) to those of lower pressure (the chest and right atrium). When the pressures are reversed during expiration (exhaling), backflow in the veins is prevented by valves.

Control of Blood Pressure

Changes in blood pressure are routinely made in order to direct appropriate amounts of oxygen and nutrients to specific parts of the body. For example, when exercise demands additional supplies of oxygen to skeletal muscles, blood delivery to these muscles increases, while blood delivery to the digestive organs decreases. Adjustments in blood pressure are also required when forces are applied to your body, such as when starting or stopping in an elevator.

Blood pressure can be adjusted by producing changes in the following variables:

- Cardiac output can be altered by changing stroke volume or heart rate.

- Resistance to blood flow in the blood vessels is most often altered by changing the diameter of the vessels (vasodilation or vasoconstriction). Changes in blood viscosity (its ability to flow) or in the length of the blood vessels (which increases with weight gain) can also alter resistance to blood flow.

The following mechanisms help regulate blood pressure:

- The cardiovascular center provides a rapid, neural mechanism for the regulation of blood pressure by managing cardiac output or by adjusting blood vessel diameter. Located in the medulla oblongata of the brain stem, it consists of three distinct regions:

■ The cardiac center stimulates cardiac output by increasing heart rate and contractility. These nerve impulses are transmitted over sympathetic cardiac nerves.

■ The cardiac center inhibits cardiac output by decreasing heart rate. These nerve impulses are transmitted over parasympathetic vagus nerves.

■ The vasomotor center regulates blood vessel diameter. Nerve impulses transmitted over sympathetic motor neurons called vasomotor nerves innervate smooth muscles in arterioles throughout the body to maintain vasomotor tone, a steady state of vasoconstriction appropriate to the region.

■ The cardiovascular center receives information about the state of the body through the following sources:

■ **Baroreceptors** are sensory neurons that monitor arterial blood pressure. Major baroreceptors are located in the carotid sinus (an enlarged area of the carotid artery just above its separation from the aorta), the aortic arch, and the right atrium.

■ **Chemoreceptors** are sensory neurons that monitor levels of CO_2 and O_2. These neurons alert the cardiovascular center when levels of O_2 drop or levels of CO_2 rise (which result in a drop in pH). Chemoreceptors are found in carotid bodies and aortic bodies located near the carotid sinus and aortic arch.

■ Higher brain regions, such as the cerebral cortex, hypothalamus, and limbic system, signal the cardiovascular center when conditions (stress, fight-or-flight response, hot or cold temperature) require adjustments to the blood pressure.

■ The kidneys provide a hormonal mechanism for the regulation of blood pressure by managing blood volume.

■ The renin-angiotensin-aldosterone system of the kidneys regulates blood volume. In response to rising blood pressure, the juxtaglomerular cells in the kidneys secrete renin into the blood. Renin converts the plasma protein angiotensinogen to angiotensin I, which in turn is converted to angiotensin II by enzymes from the lungs. Angiotensin II activates two mechanisms that raise blood pressure:

Angiotensin II constricts blood vessels throughout the body (raising blood pressure by increasing resistance to blood flow). Constricted blood vessels reduce the amount of blood delivered to the kidneys, which decreases the kidneys' potential to excrete water (raising blood pressure by increasing blood volume).

Angiotensin II stimulates the adrenal cortex to secrete aldosterone, a hormone that reduces urine output by increasing retention of H_2O and Na^+ by the kidneys (raising blood pressure by increasing blood volume).

Various substances influence blood pressure. Some important examples follow:

- Epinephrine and norepinephrine, hormones secreted by the adrenal medulla, raise blood pressure by increasing heart rate and the contractility of the heart muscles and by causing vasoconstriction of arteries and veins. These hormones are secreted as part of the fight-or-flight response.

- Antidiuretic hormone (ADH), a hormone produced by the hypothalamus and released by the posterior pituitary, raises blood pressure by stimulating the kidneys to retain H_2O (raising blood pressure by increasing blood volume).

- Atrial natriuretic peptide (ANP), a hormone secreted by the atria of the heart, lowers blood pressure by causing vasodilation and by stimulating the kidneys to excrete more water and Na^+ (lowering blood pressure by reducing blood volume).

- Nitric oxide (NO), secreted by endothelial cells, causes vasodilation.

- Nicotine in tobacco raises blood pressure by stimulating sympathetic neurons to increase vasoconstriction and by stimulating the adrenal medulla to increase secretion of epinephrine and norepinephrine.

- Alcohol lowers blood pressure by inhibiting the vasomotor center (causing vasodilation) and by inhibiting the release of ADH (increasing H_2O output, which decreases blood volume).

Blood Vessels of the Body

Figures 14-5 and 14-6 show the major arteries and veins of the body.

Figure 14-5 The major arteries in the body.

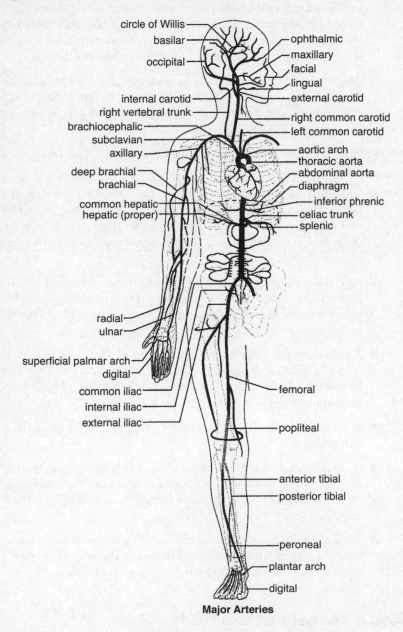

circle of Willis
basilar
occipital
ophthalmic
maxillary
facial
lingual
internal carotid
right vertebral trunk
brachiocephalic
subclavian
axillary
external carotid
right common carotid
left common carotid
aortic arch
thoracic aorta
abdominal aorta
diaphragm
deep brachial
brachial
inferior phrenic
common hepatic
hepatic (proper)
celiac trunk
splenic
radial
ulnar
superficial palmar arch
digital
common iliac
internal iliac
external iliac
femoral
popliteal
anterior tibial
posterior tibial
peroneal
plantar arch
digital

Major Arteries

Figure 14-6 The major veins in the body.

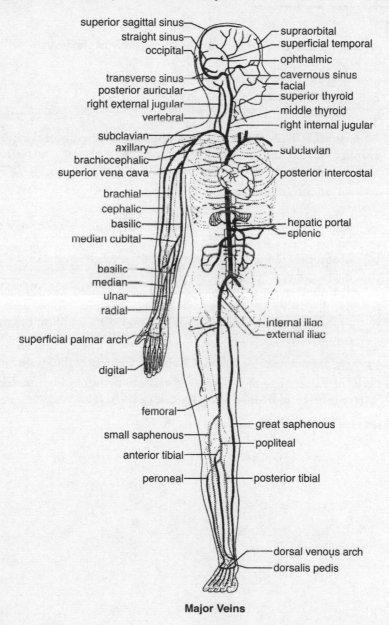

superior sagittal sinus
straight sinus
occipital

transverse sinus
posterior auricular
right external jugular
vertebral

subclavian
axillary
brachiocephalic
superior vena cava

brachial
cephalic
basilic
median cubital

basilic
median
ulnar
radial

superficial palmar arch

digital

femoral

small saphenous

anterior tibial

peroneal

supraorbital
superficial temporal
ophthalmic
cavernous sinus
facial
superior thyroid
middle thyroid
right internal jugular

subclavian

posterior intercostal

hepatic portal
splenic

internal iliac
external iliac

great saphenous
popliteal

posterior tibial

dorsal venous arch
dorsalis pedis

Major Veins

Chapter Check-Out

Q&A

1. Which variable is not a measure of the capacity of the heart?

    a. Cardiac output (CO)
    b. Stroke volume (SV)
    c. Heart rate (HR)
    d. End-systolic volume (ESV)

2. True or False: The cardiac cycle describes all the activities of the heart through one complete heartbeat—that is, through one contraction and relaxation of both the atria and ventricles.

3. Which of the following is not a function of the cardiovascular system?

    a. Regulation of body temperature, fluid pH, and water content of cells
    b. Absorption of lipids and fat-soluble materials from the digestive tract
    c. Transport of nutrients, oxygen, and hormones to cells throughout the body
    d. Protection of the body by white blood cells, antibodies, and complement proteins

4. _____ are microscopic blood vessels with extremely thin walls.

5. True or False: Alcohol raises blood pressure by inhibiting the vasomotor center and by inhibiting the release of ADH.

Answers: 1. c, **2.** T, **3.** b, **4.** Capillaries, **5.** F

Chapter 15

THE LYMPHATIC SYSTEM

An important supplement to the cardiovascular system in helping to remove toxins from the body, the lymphatic system is also a crucial support of the immune system. Unlike blood, lymph only moves one way through your body, propelled by the action of nearby skeletal muscles. The lymph is pushed into the bloodstream for elimination. Appreciating the importance of the lymphatic system in filtering, recycling, and producing blood as well as filtering lymph, collecting excess fluids, and absorbing fat-soluble materials is necessary to the understanding of human physiology.

Lymphatic System Components

The lymphatic system consists of lymphatic vessels, a fluid called **lymph,** lymph nodes, the thymus, and the spleen. This system supplements and extends the cardiovascular system in the following ways:

■ The lymphatic system collects excess fluids and plasma proteins from surrounding tissues (interstitial fluids) and returns them to the blood circulation. Because lymphatic capillaries are more porous than blood capillaries, they are able to collect fluids, plasma proteins, and blood cells that have escaped from the blood. Within lymphatic vessels, this collected material forms a usually colorless fluid called lymph, which is transported to the right and left subclavian veins of the circulatory system.

■ The lymphatic system absorbs lipids and fat-soluble materials from the digestive tract.

■ The lymphatic system filters the lymph by destroying pathogens, inactivating toxins, and removing particulate matter. Lymph nodes, small bodies interspersed along lymphatic vessels, act as cleaning filters and as immune response centers that defend against infection.

The movement of lymph through lymphatic vessels is slow (3 liters/day) compared to blood flow (about 5 liters/minute). Lymph does not circulate like blood, but moves in one direction from its collection in tissues to its return in the blood. There are no lymphatic pumps. Instead, lymph, much like blood in veins, is propelled forward by the action of the nearby skeletal muscles, the expansion and contraction of the lungs, and the contraction of the smooth muscle fibers in the walls of the lymphatic vessels. Valves in the lymphatic vessels prevent the backward movement of lymph.

Lymphatic Vessels

Lymphatic vessels occur throughout the body alongside arteries (in the viscera) or veins (in the subcutaneous tissue). They are absent from the central nervous system, bone marrow, teeth, and avascular tissues.

■ Lymph capillaries, the smallest lymphatic vessels, begin as dead-end vessels. They resemble blood capillaries, but are much more porous to surrounding fluids due to the following two features:

- Valvelike openings form at the juncture of adjacent endothelial cells. Unlike the tightly joined endothelial cells that make up the walls of blood capillaries, those of lymph capillaries loosely overlap. When fluid pressure increases in surrounding regions, the overlapped cells separate, allowing fluids to enter the lymph capillary. When pressure inside the capillary exceeds the pressure outside, the spaces between the endothelial cells close, holding fluids inside the capillary.

- Anchoring filaments attach the endothelial cells of the lymphatic vessels to surrounding collagen. When interstitial fluid pressure increases, the anchoring filaments prevent the endothelial cells from collapsing, keeping the spaces between the endothelial cells open.

- Lacteals are specialized lymph capillaries that occur in the fingerlike projections (villi) that extend into the small intestine. Lacteals absorb lipids from the intestinal tract. The lymph within these capillaries, called chyle, has a creamy white color (rather than clear) due to the presence of fats.

- Lymphatic collecting vessels form as lymph capillaries merge. Collecting vessels have the following characteristics:

  - Valves are present to prevent the backward flow of lymph (as in veins).

  - The walls of collecting vessels consist of the same three tunics (layers) that characterize veins, but the layers are thinner and poorly defined.

- Lymphatic trunks form from the union of collecting vessels. The nine major trunks, draining lymph from regions for which they are named, are the lumbar, jugular, subclavian, and bronchomediastinal trunks, each of which occurs in pairs (left and right, for each side of the body), and a single intestinal trunk.

- Lymphatic ducts are the largest lymphatic vessels. These two ducts drain lymph into veins in the neck (the right and left subclavian veins at their junctures with the internal jugular veins). Valves in the lymphatic ducts at their junctures with the veins prevent the entrance of blood into the lymphatic vessels.

■ The thoracic duct collects lymph from the left side of the body and regions of the right side of the body below the thorax. It ultimately drains lymph into the left subclavian vein. It begins at the cisterna chili, an enlarged region of the lymphatic vessel that forms following the union of the intestinal trunk and right and left lumbar trunks.

■ The right thoracic duct collects lymph from the upper right side of the body (right arm and right regions of thorax, neck, and head), a much smaller area than that serviced by the thoracic duct. It ultimately drains lymph into the right subclavian vein.

Figure 15-1 illustrates the location of the thoracic duct and the left thoracic duct.

Figure 15-1 Lymphatic trunks and ducts.

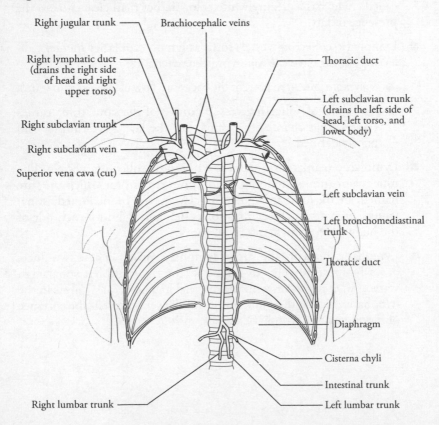

Lymphoid Cells

Lymphatic (lymphoid) tissue is a kind of connective tissue. It consists of the following types of cells:

■ Lymphocytes are white blood cells (**leukocytes**) that provide an immune response that attacks specific kinds of nonself cells and foreign substances (antigens). There are several major classes of lymphocytes (two are mentioned here; more are discussed in Chapter 16):

■ **T cells** (T lymphocytes) originate in the bone marrow but mature in the thymus gland. T cells attack self cells that have been invaded by pathogens, abnormal self cells (such as cancerous cells), or nonself cells (such as those that might be introduced in an organ transplant).

■ **B cells** (B lymphocytes) originate and mature in the bone marrow. When B cells encounter an antigen (a toxin, virus, or bacterium), they produce plasma cells and memory cells. Plasma cells release antibodies that bind to the antigen and inactivate it. Memory cells circulate in the lymph and blood with the capacity to produce additional antigens for future encounters with the same antigen.

■ Macrophages are enlarged monocytes (white blood cells) that engulf microbes and cellular debris.

■ Reticular cells and their reticular fibers made from collagen and glycoproteins provide a network within which the lymphocytes and other cells reside.

Lymphatic Tissues and Organs

Lymphatic cells are organized into tissues and organs based on how tightly the lymphatic cells are arranged and whether the tissue is encapsulated by a layer of connective tissue. Three general categories exist:

■ Diffuse, unencapsulated bundles of lymphatic cells. This kind of lymphatic tissue consists of lymphocytes and macrophages associated with a reticular fiber network. It occurs in the lamina propria (middle layer) of the mucus membranes (mucosae) that line the respiratory and gastrointestinal tracts.

- Discrete, unencapsulated bundles of lymphatic cells, called lymphatic nodules (follicles). These bundles have clear boundaries that separate them from neighboring cells. Nodules occur within the lamina propria of the mucus membranes that line the gastrointestinal, respiratory, reproductive, and urinary tracts. They are referred to as **mucosa-associated lymphoid tissue (MALT)**. The nodules contain lymphocytes and macrophages that protect against bacteria and other pathogens that may enter these passages with food, air, or urine. Nodules occur as solitary nodules, or they cluster as patches or aggregates. Here are the major clusters of nodules:

  - **Peyer's patches** are clusters of lymphatic nodules that occur in the mucosa that lines the ileum of the small intestine.

  - The **tonsils** are aggregates of lymphatic nodules that occur in the mucosa that lines the pharynx (throat). Each of the seven tonsils that form a ring around the pharynx are named for their specific region: a single pharyngeal tonsil (**adenoid**) in the rear wall of the nasopharynx, two palatine tonsils on each side wall of the oral cavity at its entrance in the throat, two lingual tonsils at the base of the tongue, and two small tubal tonsils in the pharynx at the entrance to the auditory tubes.

  - The appendix, a small fingerlike attachment to the beginning of the large intestine, is lined with aggregates of lymph nodules.

- Encapsulated organs contain lymphatic nodules and diffuse lymphatic cells surrounded by a capsule of dense connective tissue. The three lymphatic organs are discussed in the following sections.

Lymph nodes

Lymph nodes are small, oval, or bean-shaped bodies that occur along lymphatic vessels. They are abundant where lymphatic vessels merge to form trunks, especially in the inguinal (groin), axillary (armpit), and mammary gland areas. Lymph flows into a node through afferent lymphatic vessels that enter the convex side of a node. It exits the node at the hilus, the indented region on the opposite, concave side of the node, through efferent lymphatic vessels. Efferent vessels contain valves that restrict lymph to movement in one direction out of the lymph node. The number of efferent vessels leaving the lymph node is fewer than the number of afferent vessels entering, slowing the flow of lymph through the node.

Lymph nodes perform three functions:

■ *They filter the lymph,* preventing the spread of microorganisms and toxins that enter interstitial fluids.

■ *They destroy bacteria, toxins, and particulate matter* through the phagocytic action of macrophages.

■ *They produce antibodies* through the activity of B cells.

The structure of a lymph node is characterized by the following features:

■ There is a capsule of dense connective tissue that surrounds the lymph node.

■ Trabeculae are projections of the capsule that extend into the node, forming compartments. The trabeculae support reticular fibers that form a network that supports lymphocytes.

■ The cortex is the dense, outer region of the node. It contains lymphatic nodules where B cells and macrophages proliferate.

■ The medulla is the center of the node. Less dense than the surrounding cortex, the medulla primarily contains T cells.

■ Medullary cords are strands of reticular fibers with lymphocytes and macrophages that extend from the cortex toward the hilus.

■ Sinuses are passageways through the cortex and medulla through which lymph moves toward the hilus.

Thymus

The thymus is a bilobed organ located in the upper chest region between the lungs, posterior to the sternum. It grows during childhood and reaches its maximum size of 40 g at puberty. It then slowly decreases in size as it is replaced by adipose and areolar connective tissue. By age 65, it weighs about 6 g.

Each lobe of the thymus is surrounded by a capsule of connective tissue. Lobules produced by trabeculae (inward extensions of the capsule) are characterized by an outer cortex and inner medulla. The following cells are present:

■ *Lymphocytes* consist almost entirely of T cells.

■ *Epithelial-reticular cells* resemble reticular cells, but do not form reticular fibers. Instead, these star-shaped cells form a reticular network by interlocking their slender cellular processes (extensions). These processes are held together by desmosomes, cell junctions formed by protein fibers. Epithelial-reticular cells produce thymosin and other hormones believed to promote the maturation of T cells.

The function of the thymus is to promote the maturation of T lymphocytes. Immature T cells migrate through the blood from the red bone marrow to the thymus. Within the thymus, the immature T cells concentrate in the cortex, where they continue their development. Mature T cells leave the thymus by way of blood vessels or efferent lymphatic vessels, migrating to other lymphatic tissues and organs where they become active (immunocompetent) in immune responses. The thymus does not provide a filtering function similar to lymph nodes (there are no afferent lymphatic vessels leading into the thymus), and unlike all other centers of lymphatic tissues, the thymus does not play a direct role in immune responses.

Blood vessels that permeate the thymus are surrounded by epithelial-reticular cells. These cells establish a protective blood-thymus barrier that prevents the entrance of antigens from the blood and into the thymus where T cells are maturing. Thus, an antigen-free environment is maintained for the development of T cells.

Spleen

Measuring about 12 cm (5 inches) in length, the spleen is the largest lymphatic organ. It is located on the left side of the body, inferior to the diaphragm and at the left edge of the stomach. Like other lymphatic organs, the spleen is surrounded by a capsule whose extensions into the spleen form trabeculae. The splenic artery, splenic vein, nerves, and efferent lymphatic vessels pass through the hilus of the spleen located on its slightly concave, upper surface. There are two distinct areas within the spleen:

■ White pulp consists of reticular fibers and lymphocytes in nodules that resemble the nodules of lymph nodes.

■ Red pulp consists of venous sinuses filled with blood. Splenic cords consisting of reticular connective tissue, macrophages, and lymphocytes form a mesh between the venous sinuses and act as a filter as blood passes between arterial vessels and the sinuses.

The functions of the spleen include the following:

■ *The spleen filters the blood.* Macrophages in the spleen remove bacteria and other pathogens, cellular debris, and aged blood cells. There are no afferent lymphatic vessels, and unlike lymph nodes, the spleen does not filter lymph.

■ *The spleen destroys old red blood cells and recycles their parts.* It removes the iron from heme groups and binds the iron to the storage protein.

■ *The spleen provides a reservoir of blood.* The diffuse nature of the red pulp retains large quantities of blood, which can be directed to the circulation when necessary. One third of the blood platelets are stored in the spleen.

■ *The spleen is active in immune responses.* T cells proliferate in the white pulp before returning to the blood to attack nonself cells when necessary. B cells proliferate in the white pulp, producing plasma cells and antibodies that return to the blood to inactivate antigens.

■ *The spleen produces blood cells.* Red and white blood cells are produced in the spleen during fetal development.

Chapter Check-Out

Q&A

1. True or False: The spleen is the largest lymphatic organ.

2. Which of the following is not a function of the lymph nodes?
   a. They filter the lymph.
   b. They destroy old red blood cells.
   c. They destroy bacteria, toxins, and particulate matter.
   d. They produce antibodies.

3. The lymphatic system filters the lymph by destroying _____, inactivating toxins, and removing particulate matter.

4. Which of the following are lymphatic vessels?
   a. Lacteals
   b. Lymphocytes
   c. Macrophages
   d. Reticular cells

5. True or False: The kind of lymphatic tissue that consists of lymphocytes and macrophages associated with a reticular fiber network is called lymphatic nodules.

Answers: 1. T, **2.** b, **3.** pathogens, **4.** a, **5.** F

Chapter 16

THE IMMUNE SYSTEM AND OTHER BODY DEFENSES

Chapter Check-In

❏ Listing nonspecific barriers and nonspecific defenses

❏ Recognizing the five main tasks of the immune system

❏ Understanding the major histocompatibility complex

❏ Describing the various kinds of lymphocytes

❏ Knowing the different immune responses and supplements to the immune response

The colossal task of keeping the body safe from outside and inside attacks of bacteria, viruses, and other nasty critters belongs to the immune system. Your skin and mucous membranes are the first line of defense from invaders entering through the skin or through openings in the body. A second line of defense exists inside the body to challenge invaders that make it through the first line of defense. Sounds more like a war strategy, doesn't it? That's how your body treats anything that isn't supposed to be there—as an invader that must be destroyed.

Protecting Your Body

The internal environment of the human body provides attractive conditions for growth of bacteria, viruses, and other organisms. Although some of these organisms can live symbiotically within humans, many either cause destruction of cells or produce toxic chemicals. To protect against these foreign invaders, three lines of defense are employed: nonspecific barriers that deter the entrance of invaders, and both nonspecific and specific defenses against invaders inside the body. A nonspecific defense is a rapid response to a wide range of pathogens. A specific defense, delivered by the immune system, takes several days to mount and target specific invaders that escape the attack of the nonspecific defense.

Nonspecific Barriers

The skin and mucous membranes provide a nonspecific first line of defense against invaders entering through the skin or through openings into the body. The first line of defense features the following mechanisms:

■ Skin is a physical and hostile barrier covered with oily and acidic (pH from 3 to 5) secretions from sebaceous and sweat glands, respectively.

■ Antimicrobial proteins (such as lysozyme, which breaks down the cell walls of bacteria) are contained in saliva, tears, and other secretions found on mucous membranes.

■ Cilia that line the respiratory tubes serve to sweep invaders away from the lungs.

■ Gastric juices of the stomach, by the action of hydrochloric acid or enzymes, kill most microbes.

■ Symbiotic bacteria found in the digestive tract and the vagina outcompete many other organisms that could cause damage.

Nonspecific Defenses

The second line of defense consists of mechanisms or agents that indiscriminately challenge foreign invaders that are inside the body:

■ **Phagocytes** are white blood cells (leukocytes) that engulf pathogens by phagocytosis. They include neutrophils, monocytes, and eosinophils. Monocytes enlarge into large phagocytic cells called macrophages.

■ Natural killer cells (NK cells) are lymphocytes (white blood cells that mature in lymphoid tissues). NK cells kill pathogen-infected body cells or abdominal body cells (such as tumors).

■ Complement proteins are a group of about 20 proteins that "complement" defense reactions. These proteins help attract phagocytes to foreign cells and help destroy foreign cells by promoting cell lysis (breaking open the cell).

■ **Interferons (IFNs)** are substances secreted by cells invaded by viruses that stimulate neighboring cells to produce proteins that help them defend against the viruses. Certain IFNs (such as gamma-IFN) also amplify the activity of macrophages and NK cells.

■ The inflammatory response is a series of nonspecific events that occur in response to pathogens. The response typically produces redness, swelling, heat, and pain in the target area, and often the area is disabled. When skin is damaged, for example, and bacteria, other organisms, or toxic substances enter the body, the following events occur:

1. A chemical alarm is generated in the injured area. Injured cells and nearby circulating cells release chemicals that initiate defensive actions and sound an alarm to other defense mechanisms. These chemicals include histamine (mostly secreted by basophils, white blood cells found in connective tissue), kinins, prostaglandins (PGs), and complement proteins.

2. **Vasodilation** (dilation of blood vessels), stimulated by histamine and other chemicals, increases blood supply to the damaged area. This causes redness and an increase in local temperature. The increase in temperature stimulates white blood cells and makes the environment inhospitable to pathogens.

3. Vascular permeability increases in response to alarm chemicals. As a result, white blood cells, clotting factors, and body fluids move more quickly through blood vessel walls and into the injured area. The increase in body fluids that results causes local edema (swelling). Edema may produce pain if nearby nerve endings experience pressure. Pain may also occur when nerve endings are exposed to bacterial toxins, kinins, and prostaglandins. (Aspirin reduces pain by inhibiting the production of prostaglandins.)

4. Phagocytes arrive at the site of injury and engulf pathogens and damaged cells. Phagocytes find the site of injury by **chemotaxis,** the movement of cells in response to chemical gradients (provided here by alarm chemicals).

5. Complement proteins help phagocytes engulf foreign cells and stimulate basophils to release histamine.

■ Fever is a total body response to infection characterized by elevated body temperature. An elevated temperature increases cellular metabolism (accelerating cellular repairs), amplifies the effect of alarm chemicals, and creates a hostile environment for bacteria. An excessively high fever may reduce the activity of enzymes necessary for cellular metabolism, thus causing the body to go further out of homeostasis.

Specific Defense (The Immune System)

The immune system is the third line of defense. It consists of mechanisms and agents that target specific antigens (Ags). An antigen is any molecule, usually a protein or polysaccharide, that can be identified as foreign (nonself) or self (such as MHC antigens described below). It may be a toxin (injected into the blood by the sting of an insect, for example), a part of the protein coat of a virus, or a molecule unique to the plasma membranes of bacteria, protozoa, pollen, or other foreign cells. Once the foreign antigen is recognized, an agent is released that targets that specific antigen. In the process of mounting a successful defense, the immune system accomplishes five tasks:

■ *Recognition.* The antigen or cell is recognized as nonself. To differentiate self from nonself, unique molecules on the plasma membrane of cells called the **major histocompatibility complex (MHC)** are used as a means of identification.

■ *Lymphocyte selection.* The primary defending cells of the immune system are certain white blood cells called lymphocytes. The immune system potentially possesses billions of lymphocytes, each equipped to target a different antigen. When an antigen, or nonself cell, binds to a lymphocyte, the lymphocyte proliferates, producing numerous daughter cells, all identical copies of the parent cell. This process is called *clonal selection* because the lymphocyte to which the antigen effectively binds is "selected" and subsequently reproduces to make clones, or identical copies, of itself.

■ *Lymphocyte activation.* The binding of an antigen or foreign cell to a lymphocyte may activate the lymphocyte and initiate proliferation. In most cases, however, a costimulator is required before proliferation begins. Costimulators may be chemicals or other cells.

■ *Destruction of the foreign substance.* Lymphocytes and antibodies destroy or immobilize the foreign substance. Nonspecific defense mechanisms (phagocytes, NK cells) help eliminate the invader.

■ *Memorization.* Long-lived "memory" lymphocytes are produced and can quickly recognize and respond to future exposures to the antigen or foreign cell.

Major Histocompatibility Complex

The major histocompatibility complex (MHC) (also called human leukocyte antigens, HLAs) is the mechanism by which the immune system is able to differentiate between self and nonself cells. The MHC is a collection of glycoproteins (proteins with a carbohydrate) that exist on the plasma membranes of nearly all body cells. The proteins of a single individual are unique, originating from 20 genes, with more than 50 variations per gene between individuals. Thus, it is extremely unlikely that two people, except for identical twins, will possess cells with the same set of MHC molecules.

The immune system is able to identify nonself cells by aberrations in the MHC displayed on the plasma membrane. There are two groups of MHC molecules, and each group generates different markings on the plasma membrane:

■ MHC-I glycoproteins are produced by all body cells (except red blood cells). When a cell becomes cancerous or is invaded by a virus, unfamiliar proteins are synthesized in the cell. These proteins are endogenous antigens—that is, antigens produced inside the cell. Portions of these antigens are combined with MHC-I glycoproteins and, when displayed on the plasma membrane, indicate a nonself cell.

■ MHC-II glycoproteins are produced only by antigen-presenting cells (APCs)—mostly macrophages and B cells. APCs actively ingest exogenous antigens—antigens that originate outside the cell. Exogenous antigens include viruses, toxins, pollen, or bacteria that are circulating in the blood, lymph, or body fluids. APCs break down the antigens and incorporate pieces of them with MHC-II glycoproteins. This aberrant display of MHC markers is recognized as nonself.

Lymphocytes

The primary agents of the immune response are lymphocytes, white blood cells (leukocytes) that originate in the bone marrow (like all blood cells) but concentrate in lymphoid tissues such as the lymph nodes, the thymus gland, and the spleen. When lymphocytes mature, they become immunocompetent, or capable of binding with a specific antigen. An immunocompetent lymphocyte displays unique proteins on its plasma membrane that act as antigen receptors. Because all of the antigen receptors of an individual lymphocyte are identical, only a specific antigen can bind to an individual lymphocyte. The kind of antigen receptors displayed by a particular lymphocyte is determined by somatic recombination, a shuffling of gene segments during lymphocyte maturation. By mixing gene segments, more than one billion different antigen receptors can be generated.

Here are the various kinds of lymphocytes:

■ **B cells** (B lymphocytes) are lymphocytes that originate and mature in the bone marrow. The antigen receptors of B cells bind to freely circulating antigens. When B cells encounter antigens that bind to their antigen binding sites, the B cells proliferate, producing two kinds of daughter cells, plasma cells and memory cells:

  ■ Plasma cells are daughter cells of B cells. Each plasma cell releases antibodies, proteins that have the same antigen binding capability as the antigen receptors of its parent B cell. Antibodies circulate through the body, binding to the specific antigens that stimulated the proliferation of plasma cells.

  ■ Memory B cells are long-lived daughter cells of B cells that, like plasma cells, produce antibodies. However, memory cells do not release their antibodies in response to the immediate antigen invasion. Instead, the memory cells circulate in the body and respond quickly to eliminate any subsequent invasion by the same antigen. This mechanism provides immunity to many diseases after the first occurrence of the disease.

■ **T cells** (T lymphocytes) are lymphocytes that originate in the bone marrow, but mature in the thymus gland. The antigen receptors of T cells bind to self cells that display foreign antigens (with MHC proteins) on their plasma membrane. When T cells bind to these aberrant self cells, they divide and produce the following kinds of daughter cells:

  ■ Cytotoxic T cells (killer T cells) are activated when they recognize antigens that are mixed with the MHC-I proteins of self cells. Following activation, cytotoxic cells proliferate and destroy the recognized cells by producing toxins that puncture them, thus causing them to lyse.

■ Helper T cells are activated when they recognize antigens that are mixed with the MHC-II proteins of self cells. Proliferation produces helper T cells that intensify antibody production of B cells. Helper T cells also secrete hormones called cytokines that stimulate the proliferation of B cells and T cells.

■ Suppressor T cells are believed to be involved in winding down a successful immune response and in preventing the attachment of uninfected self cells.

■ Memory T cells are long-lived cells possessing the same antigen receptors as their parent T cell. Like memory B cells, they provide a rapid defense to any subsequent invasion by the same antigen.

Antibodies

Antibodies are proteins that bind to specific antigens. B cells, located in lymphoid tissue, release the antibodies, which then circulate in the blood plasma, lymph, or extracellular fluids. Some antibodies migrate to other areas of the body, such as the respiratory tract or the placenta, or enter various body secretions, such as saliva, sweat, and milk. Additional properties of antibodies include the following:

■ There are five classes of antibodies (or immunoglobulins): IgA, IgD, IgE, IgG, and IgM. Antibodies circulating in the blood are primarily IgG, IgA, and IgM; IgD and a second form of IgM antibodies are found on the plasma membranes of B cells, where they act as antigen receptors. IgE antibodies attach to basophils and mast cells (both white blood cells found in connective tissue) and induce them to secrete histamine.

■ The basic structure of an antibody is Y-shaped protein that consists of constant and variable regions. The variable regions are sequences of amino acids that differ among antibodies and give them specificity to antigens.

■ Antibodies bind to antigens, thus forming an antigen-antibody complex. This complex attracts macrophages, which will phagocytize any foreign substance that has that specific antigen-antibody complex. The formation of these complexes may also cause agglutination (clumping) of antigens or foreign cells.

Costimulation

In some immune responses, a B cell or T cell becomes activated when an antigen or nonself cell binds to it. Activation then initiates proliferation. In most immune responses, however, activation requires the presence of a costimulator. Those two signals, an antigen and a costimulator, are required to initiate the immune response, ensuring that healthy self cells are not destroyed. Costimulation may occur in two ways:

- Cytokines, released by helper T cells and APCs, act as costimulators. Cytokines are protein hormones that influence cell growth. When a helper T cell becomes activated or an APC engulfs an antigen, the helper T cell or APC secretes a cytokine called interleukin.

- Helper T cells and APCs act as costimulators. Activated T cells or APCs that display antigens activate B cells or T cells when they temporarily bind to them.

Humoral and Cell-Mediated Immune Responses

The immune system distinguishes two groups of foreign substances. One group consists of antigens that are freely circulating in the body. These include molecules, viruses, and foreign cells. A second group consists of self cells that display aberrant MHC proteins. Aberrant MHC proteins can originate from antigens that have been engulfed and broken down (exogenous antigens) or from virus-infected and tumor cells that are actively synthesizing foreign proteins (endogenous antigens). Depending on the kind of foreign invasion, two different immune responses occur:

- The humoral response (or antibody-mediated response) involves B cells that recognize antigens or pathogens that are circulating in the lymph or blood ("humor" is a medieval term for body fluid). The response follows this chain of events:

    1. Antigens bind to B cells.

    2. Interleukins or helper T cells costimulate B cells. In most cases, both an antigen and a costimulator are required to activate a B cell and initiate B cell proliferation.

3. B cells proliferate and produce plasma cells. The plasma cells bear antibodies with the identical antigen specificity as the antigen receptors of the activated B cells. The antibodies are released and circulate through the body, binding to antigens.

4. B cells produce memory cells. Memory cells provide future immunity.

■ The cell-mediated response involves mostly T cells and responds to any cell that displays aberrant MHC markers, including cells invaded by pathogens, tumor cells, or transplanted cells. The following chain of events describes this immune response:

1. Self cells or APCs displaying foreign antigens bind to T cells.

2. Interleukins (secreted by APCs or helper T cells) costimulate activation of T cells.

3. If MHC-I and endogenous antigens are displayed on the plasma membrane, T cells proliferate, producing cytotoxic T cells. Cytotoxic T cells destroy cells displaying the antigens.

4. If MHC-II and exogenous antigens are displayed on the plasma membrane, T cells proliferate, producing helper T cells. Helper T cells release interleukins (and other cytokines), which stimulate B cells to produce antibodies that bind to the antigens and stimulate nonspecific agents (NK and macrophages) to destroy the antigens.

Supplements to the Immune Response

Three important agents are used in medicine to supplement the immune response:

■ **Antibiotics** are chemicals that are harmful to bacteria.

■ **Vaccines** are substances that stimulate the production of memory cells. Inactivated viruses or fragments of viruses, bacteria, or other microorganisms are used as vaccines. Once memory cells are formed, the introduction of the live microorganism will stimulate a swift response by the immune system before any disease can become established.

■ **Passive immunity** is obtained by transferring antibodies from an individual who previously had a disease to a newly infected individual. Newborn infants are protected by passive immunity through the transfer of antibodies across the placenta and by antibodies in breast milk.

Chapter Check-Out

Q&A

1. True or False: Three important agents are used in medicine to supplement the immune response. These are antibiotics, vaccines, and passive immunity.

2. The skin and _____ _____ provide a _____ first line of defense against invaders entering through the skin or through openings into the body.

3. Which of the following is not a second line of defense that challenges foreign invaders that are inside the body?
   a. Fever
   b. Complement proteins
   c. Antimicrobial proteins
   d. Inflammatory response

4. True or False: The major histocompatibility complex (MHC) is the mechanism by which the immune system is able to differentiate between self and nonself cells.

5. Which of the following lymphocytes originate and mature in the bone marrow?
   a. T cells
   b. B cells

Answers: 1. T, **2.** mucous membranes, nonspecific, **3.** c, **4.** T, **5.** b

Chapter 17

THE RESPIRATORY SYSTEM

Chapter Check-In

❑ Describing the respiration processes

❑ Recognizing the main structures of the respiratory system

❑ Knowing Boyle's Law on the mechanics of breathing

❑ Understanding gas exchange and transport

❑ Listing the respiratory centers

Out with the old and in with the new—that's what the respiratory system does, delivering air to the lungs, bringing oxygen into the body, and expelling the carbon dioxide back into the air. Understanding the structure and intricacies of the respiratory system is vital to human anatomy. The respiratory system is made up of more than just the lungs; it also includes your nose, throat, larynx, windpipe, bronchi, alveolar ducts, and respiratory membrane.

Function of the Respiratory System

The function of the respiratory system is to deliver air to the lungs. Oxygen in the air diffuses out of the lungs and into the blood, while carbon dioxide diffuses in the opposite direction, out of the blood and into the lungs. Respiration includes the following processes:

■ External respiration is the process of gas exchange between the atmosphere and the body tissues. In order to accomplish this task, the following events occur:

1. Pulmonary ventilation is the process of breathing—inspiration (inhaling air) and expiration (exhaling air).

2. Gas transport, carried out by the cardiovascular system, is the process of distributing the oxygen throughout the body and collecting CO_2 and returning it to the lungs.

■ Internal respiration is the process of gas exchange between the blood, the interstitial fluids (fluids surrounding the cells), and the cells. Inside the cell, cellular respiration generates energy (ATP), using O_2 and glucose and producing waste CO_2.

Structure of the Respiratory System

The respiratory system is represented by the following structures, shown in Figure 17-1:

Figure 17-1 A view of the entire respiratory system and the
upper respiratory tract.

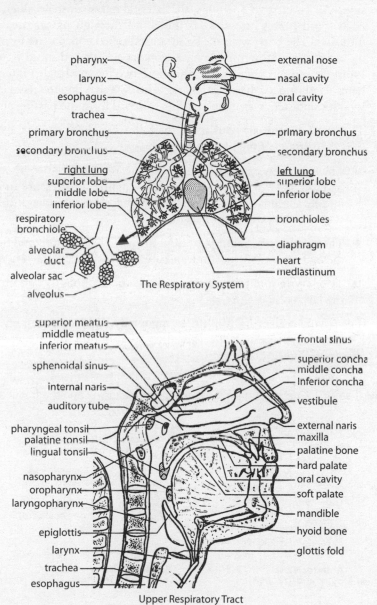

The Respiratory System

Upper Respiratory Tract

- The nose consists of the visible external nose and the internal nasal cavity. The nasal septum divides the nasal cavity into right and left sides. Air enters two openings, the external nares (nostrils; singular, naris), and passes into the vestibule and through passages called meatuses. The bony walls of the meatuses, called concha, are formed by facial bones (the inferior nasal concha and the ethmoid bone). From the meatuses, air then funnels into two (left and right) internal nares. Hair, mucus, blood capillaries, and cilia that line the nasal cavity filter, moisten, warm, and eliminate debris from the passing air.

- The pharynx (throat) consists of the following three regions, listed in order through which incoming air passes:

  - The nasopharynx receives the incoming air from the two internal nares. The two auditory tubes that equalize air pressure in the middle ear also enter here. The pharyngeal tonsil (adenoid) lies at the back of the nasopharynx.

  - The oropharyrnx receives air from the nasopharynx and food from the oral cavity. The palatine and lingual tonsils are located here.

  - The laryngopharynx passes food to the esophagus and air to the larynx.

- The larynx receives air from the laryngopharynx. It consists of several pieces of cartilage that are joined by membranes and ligaments, shown in Figure 17-2:

Figure 17-2 Anterior and sagittal sections of the larynx and the trachea.

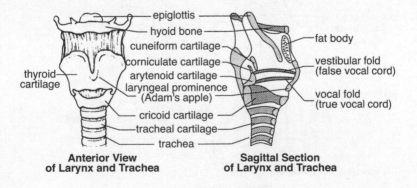

Anterior View of Larynx and Trachea

- epiglottis
- hyoid bone
- cuneiform cartilage
- corniculate cartilage
- arytenoid cartilage
- thyroid cartilage
- laryngeal prominence (Adam's apple)
- cricoid cartilage
- tracheal cartilage
- trachea

Sagittal Section of Larynx and Trachea

- fat body
- vestibular fold (false vocal cord)
- vocal fold (true vocal cord)

■ The epiglottis, the first piece of cartilage of the larynx, is a flexible flap that covers the glottis, the upper region of the larynx, during swallowing to prevent the entrance of food.

■ The thyroid cartilage protects the front of the larynx. A forward projection of this cartilage appears as the Adam's apple (anatomically known as the laryngeal prominence).

■ The paired arytenoid cartilages in the rear are horizontally attached to the thyroid cartilage in the front by folds of mucous membranes. The upper vestibular folds (false vocal cords) contain muscle fibers that bring the folds together and allow the breath to be held during periods of muscular pressure on the thoracic cavity (straining while defecating or lifting a heavy object, for example). The lower vocal folds (true vocal cords) contain elastic ligaments that vibrate when skeletal muscles move them into the path of outgoing air. Various sounds, including speech, are produced in this manner.

■ The cricoid cartilage, the paired cuneiform cartilages, and the paired corniculate cartilages are the remaining cartilages supporting the larynx.

■ The trachea (windpipe) is a flexible tube, 10 to 12 cm (4 inches) long and 2.5 cm (1 inch) in diameter (Figure 17-2.)

■ The mucosa is the inner layer of the trachea. It contains mucus-producing goblet cells and pseudostratified ciliated epithelium. The movement of the cilia sweeps debris away from the lungs toward the pharynx.

■ The submucosa is a layer of areolar connective tissue that surrounds the mucosa.

■ Hyaline cartilage forms 16 to 20 C-shaped rings that wrap around the submucosa. The rigid rings prevent the trachea from collapsing during inspiration.

■ The adventitia is the outermost layer of the trachea. It consists of areolar connective tissue.

■ The primary bronchi are two tubes that branch from the trachea to the left and right lungs.

■ Inside the lungs, each primary bronchus divides repeatedly into branches of smaller diameters, forming secondary (lobar) bronchi, tertiary (segmental) bronchi, and numerous orders of bronchioles (1 mm or less in diameter), including terminal bronchioles (0.5 mm in diameter) and microscopic respiratory bronchioles. The wall of the primary bronchi is constructed like the trachea, but as the branches of the tree get smaller, the cartilaginous rings and the mucosa are replaced by smooth muscle.

■ Alveolar ducts are the final branches of the bronchial tree. Each alveolar duct has enlarged, bubblelike swellings along its length. Each swelling is called an alveolus. Some adjacent alveoli are connected by alveolar pores.

■ The respiratory membrane consists of the alveolar and capillary walls. Gas exchange occurs across this membrane. Characteristics of this membrane follow:

■ Type I cells are thin, squamous epithelial cells that constitute the primary cell type of the alveolar wall. Oxygen diffusion occurs across these cells.

■ Type II cells are cuboidal epithelial cells that are interspersed among the type I cells. Type II cells secrete pulmonary surfactant (a phospholipid bound to a protein) that reduces the surface tension of the moisture that covers the alveolar walls. A reduction in surface tension permits oxygen to diffuse more easily into the moisture. A lower surface tension also prevents the moisture on opposite walls of an alveolus or alveolar duct from cohering and causing the minute airway to collapse.

■ Alveolar macrophage cells (dust cells) wander among the other cells of the alveolar wall, removing debris and microorganisms.

■ A thin epithelial basement membrane forms the outer layer of the alveolar wall.

■ A dense network of capillaries surrounds each alveolus. The capillary walls consist of endothelial cells surrounded by a thin basement membrane. The basement membranes of the alveolus and the capillary are often so close that they fuse.

Lungs

The lungs are a pair of cone-shaped bodies that occupy the thorax (refer to Figure 17-1). The mediastinum, the cavity containing the heart, separates the two lungs. The left and right lungs are divided by fissures into two and three lobes, respectively. Each lobe of the lung is further divided into bronchopulmonary segments (each with a tertiary bronchus), which are further divided into lobules (each with a terminal bronchiole). Blood vessels, lymphatic vessels, and nerves penetrate each lobe.

Each lung has the following superficial features:

- The apex and base identify the top and bottom of the lung, respectively.

- The costal surface of each lung borders the ribs (front and back).

- On the medial (mediastinal) surface, where each lung faces the other lung, the bronchi, blood vessels, and lymphatic vessels enter the lung at the hilus.

The pleura is a double-layered membrane consisting of an inner pulmonary (visceral) pleura, which surrounds each lung, and an outer parietal pleura, which lines the thoracic cavity. The narrow space between the two membranes, the pleural cavity, is filled with pleural fluid, a lubricant secreted by the pleura.

Mechanics of Breathing

Boyle's Law describes the relationship between the pressure (P) and the volume (V) of a gas. The law states that if the volume increases, then the pressure must decrease (or vice versa). This relationship is often written algebraically as PV = constant, or $P_1V_1 = P_2V_2$. Both equations state that the product of the pressure and volume remains the same. (Boyle's Law applies only when the temperature does not change.)

Breathing occurs when the contraction or relaxation of muscles around the lungs changes the total volume of air within the air passages (bronchi, bronchioles) inside the lungs. When the volume of the lungs changes, the pressure of the air in the lungs changes in accordance with Boyle's Law. If the pressure is greater in the lungs than outside the lungs, then air rushes out. If the opposite occurs, then air rushes in. Here is a summary of the process:

1. Inspiration occurs when the inspiratory muscles—that is, the diaphragm and the external intercostal muscles—contract. Contraction of the diaphragm (the skeletal muscle below the lungs) causes an increase in the size of the thoracic cavity, while contraction of the external intercostal muscles elevates the ribs and sternum. Thus, both muscles cause the lungs to expand, increasing the volume of their internal air passages. In response, the air pressure inside the lungs decreases below that of air outside the body. Because gases move from regions of high pressure to low pressure, air rushes into the lungs.

2. Expiration occurs when the diaphragm and external intercostal muscles relax. In response, the elastic fibers in lung tissue cause the lungs to recoil to their original volume. The pressure of the air inside the lungs then increases above the air pressure outside the body, and air rushes out. During high rates of ventilation, expiration is facilitated by contraction of the expiratory muscles (the intercostal muscles and the abdominal muscles).

Lung compliance is a measure of the ability of the lungs and thoracic cavity to expand. Due to the elasticity of lung tissue and the low surface tension of the moisture in the lungs (from the surfactant), the lungs normally have high compliance.

Lung Volumes and Capacities

The following terms describe the various lung (respiratory) volumes:

■ The **tidal volume (TV),** about 500 mL, is the amount of air inspired during normal, relaxed breathing.

■ The **inspiratory reserve volume (IRV),** about 3,100 mL, is the additional air that can be forcibly inhaled after the inspiration of a normal tidal volume.

■ The **expiratory reserve volume (ERV),** about 1,200 mL, is the additional air that can be forcibly exhaled after the expiration of a normal tidal volume.

■ **Residual volume (RV),** about 1,200 mL, is the volume of air still remaining in the lungs after the expiratory reserve volume is exhaled.

Summing specific lung volumes produces the following lung capacities.

- The **total lung capacity (TLC),** about 6,000 mL, is the maximum amount of air that can fill the lungs (TLC = TV + IRV + ERV + RV).

- The **vital capacity (VC),** about 4,800 mL, is the total amount of air that can be expired after fully inhaling (VC = TV + IRV + ERV = approximately 80 percent TLC). The value varies according to age and body size.

- The **inspiratory capacity (IC),** about 3,600 mL, is the maximum amount of air that can be inspired (IC = TV + IRV).

- The **functional residual capacity (FRC),** about 2,400 mL, is the amount of air remaining in the lungs after a normal expiration (FRC = RV + ERV).

Some of the air in the lungs does not participate in gas exchange. Such air is located in the anatomical dead space within bronchi and bronchioles—that is, outside the alveoli.

Gas Exchange

In a mixture of different gases, each gas contributes to the total pressure of the mixture. The contribution of each gas, called the partial pressure, is equal to the pressure that the gas would have if it were alone in the enclosure. **Dalton's Law** states that the sum of the partial pressures of each gas in a mixture is equal to the total pressure of the mixture.

The following factors determine the degree to which a gas will dissolve in a liquid:

- *The partial pressure of the gas.* According to **Henry's Law,** the greater the partial pressure of a gas, the greater the diffusion of the gas into the liquid.

- *The solubility of the gas.* The ability of a gas to dissolve in a liquid varies with the kind of gas and the liquid.

- *The temperature of the liquid.* Solubility decreases with increasing temperature.

Gas exchange occurs in the lungs between alveoli and blood plasma and throughout the body between plasma and interstitial fluids. The following factors facilitate diffusion of O_2 and CO_2 at these sites:

- *Partial pressures and solubilities.* Poor solubility can be offset by a high partial pressure (or vice versa). Compare the following characteristics of O_2 and CO_2:

  - Oxygen. The partial pressure of O_2 in the lungs is high (air is 21 percent O_2), but it has poor solubility properties.

  - Carbon dioxide. The partial pressure of CO_2 in air is extremely low (air is only 0.04 percent CO_2), but its solubility in plasma is about 24 times that of O_2.

- *Partial pressure gradients.* A gradient is a change in some quantity from one region to another. Diffusion of a gas into a liquid (or the reverse) occurs down a partial pressure gradient—that is, from a region of higher partial pressure to a region of lower partial pressure. For example, the strong partial pressure gradient for O_2 (pO_2) from alveoli to deoxygenated blood (105 mm Hg in alveoli versus 40 mm Hg in blood) facilitates rapid diffusion.

- *Surface area for gas exchange.* The expansive surface area of the lungs promotes extensive diffusion.

- *Diffusion distance.* Thin alveolar and capillary walls increase the rate of diffusion.

Gas Transport

Oxygen is transported in the blood in two ways:

- A small amount of O_2 (1.5 percent) is carried in the plasma as a dissolved gas.

- Most oxygen (98.5 percent) carried in the blood is bound to the protein hemoglobin in red blood cells. A fully saturated oxyhemoglobin (HbO_2) has four O_2 molecules attached. Without oxygen, the molecule is referred to as deoxyhemoglobin (Hb).

The ability of hemoglobin to bind to O_2 is influenced by the partial pressure of oxygen. The greater the partial pressure of oxygen in the blood, the more readily oxygen binds to Hb. The oxygen-hemoglobin dissociation curve, shown in Figure 17-3, shows that as pO_2 increases toward 100 mm Hg, Hb saturation approaches 100 percent. The following four factors decrease the affinity, or strength of attraction, of Hb for O_2 and result in a shift of the O_2-Hb dissociation curve to the right:

- Increase in temperature.

- Increase in partial pressure of CO_2 (pCO_2).

- Increase in acidity (decrease in pH). The decrease in affinity of Hb for O_2, called the Bohr effect, results when H^+ binds to Hb.

- Increase in BPG (bisphosphoglycerate) in red blood cells. BPG is generated in red blood cells when they produce energy from glucose.

Figure 17-3 The oxygen-hemoglobin dissociation curve.

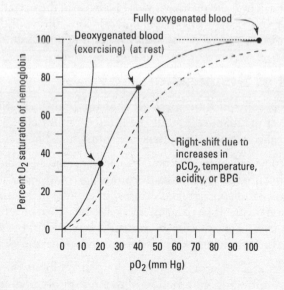

Carbon dioxide is transported in the blood in the following ways:

- A small amount of CO_2 (5 percent) is carried in the plasma as a dissolved gas.

- Some CO_2 (10 percent) binds to Hb in red blood cells, forming carbaminohemoglobin ($HbCO_2$). (The CO_2 binds to the amino acid portion of hemoglobin instead of to the iron portion.)

- Most CO_2 (85 percent) is transported as dissolved bicarbonate ions (HCO_3^-) in the plasma. The formation of HCO_3^-, however, occurs in the red blood cells, where the formation of carbonic acid (H_2CO_3) is catalyzed by the enzyme carbonic anhydrase, as follows:

$$CO_2 + H_2O \leftarrow \rightarrow H_2CO_3 \leftarrow \rightarrow H^+ + HCO_3^-$$

Following their formation in the red blood cells, most H^+ bind to hemoglobin molecules (causing the Bohr effect) while the remaining H^+ diffuse back into the plasma, slightly decreasing the pH of the plasma. The HCO_3^- ions diffuse back into the plasma as well. To balance the overall increase in negative charges entering the plasma, chloride ions diffuse in the opposite direction, from the plasma to the red blood cells (chloride shift).

Control of Respiration

Respiration is controlled by these areas of the brain that stimulate the contraction of the diaphragm and the intercostal muscles. These areas, collectively called respiratory centers, are summarized here:

- The medullary inspiratory center, located in the medulla oblongata, generates rhythmic nerve impulses that stimulate contraction of the inspiratory muscles (diaphragm and external intercostal muscles). Normally, expiration occurs when these muscles relax, but when breathing is rapid, the inspiratory center facilitates expiration by stimulating the expiratory muscles (internal intercostal muscles and abdominal muscles).

- The pheumotaxic area, located in the pons, inhibits the inspiratory center, limiting the contraction of the inspiratory muscles, and preventing the lungs from overinflating.

- The apneustic area, also located in the pons, stimulates the inspiratory center, prolonging the contraction of inspiratory muscles.

The respiratory centers are influenced by stimuli received from the following three groups of sensory neurons:

■ Central chemoreceptors (nerves of the central nervous system), located in the medulla oblongata, monitor the chemistry of cerebrospinal fluid. When CO_2 from the plasma enters the cerebrospinal fluid, it forms HCO_3^- and H^+, and the pH of the fluid drops (becomes more acidic). In response to the decrease in pH, the central chemoreceptors stimulate the respiratory center to increase the inspiratory rate.

■ Peripheral chemoreceptors (nerves of the peripheral nervous system), located in aortic bodies in the wall of the aortic arch and in carotid bodies in the walls of the carotid arteries, monitor the chemistry of the blood. An increase in pH or pCO_2, or a decrease in pO_2, causes these receptors to stimulate the respiratory center.

■ Stretch receptors in the walls of bronchi and bronchioles are activated when the lungs expand to their physical limit. These receptors signal the respiratory center to discontinue stimulation of the inspiratory muscles, allowing expiration to begin. This response is called the **inflation (Hering-Breur) reflex.**

Chapter Check-Out

Q&A

1. Which of the following is not a process of respiration?

   a. Secretion

   b. Pulmonary ventilation

   c. External respiration

   d. Gas transport

2. True or False: Boyle's Law states that if the volume increases, then the pressure must increase.

3. Most oxygen carried in the blood is bound to the protein _____.

4. Which of the following is not a respiratory center of the brain?

   a. Pneumonic area

   b. Stretch receptors

   c. Medullary inspiratory center

   d. Apneustic area

5. The _____ _____ are two tubes that branch from the trachea to the left and right lungs.

Answers: **1.** a, **2.** F, **3.** hemoglobin, **4.** b, **5.** primary bronchi

Chapter 18

THE DIGESTIVE SYSTEM

Chapter Check-In

❏ Describing the processes involved in the treatment of food

❏ Understanding the structure of the gastrointestinal tract wall

❏ Summarizing the groups of molecules encountered during digestion

❏ Knowing the basic parts of the digestive system and their main functions

❏ Listing important hormones and their effects on the regulation of digestion

■f you've ever wondered what happens to food after you eat it, this is the chapter where you find out that information. Your digestive system is made up of two major parts and it has two main functions: digestion and absorption. This complex system of breaking down food and absorbing the vital nutrients is something you should have a good understanding of in order to understand human physiology.

Function of the Digestive System

The function of the digestive system is digestion and absorption. Digestion is the breakdown of food into small molecules, which are then absorbed into the body. The digestive system is divided into two major parts:

■ The digestive tract (alimentary canal) is a continuous tube with two openings: the mouth and the anus. It includes the mouth, pharynx, esophagus, stomach, small intestine, and large intestine. Food passing through the internal cavity, or lumen, of the digestive tract does not technically enter the body until it is absorbed through the walls of the digestive tract and passes into blood or lymphatic vessels.

■ Accessory organs include the teeth and tongue, salivary glands, liver, gallbladder, and pancreas.

The treatment of food in the digestive system involves the following seven processes:

1. *Ingestion* is the process of eating.

2. *Propulsion* is the movement of food along the digestive tract. The major means of propulsion is peristalsis, a series of alternating contractions and relaxations of smooth muscle that lines the walls of the digestive organs and that forces food to move forward.

3. *Secretion* of digestive enzymes and other substances liquefies, adjusts the pH of, and chemically breaks down the food.

4. *Mechanical digestion* is the process of physically breaking down food into smaller pieces. This process begins with the chewing of food and continues with the muscular churning of the stomach. Additional churning occurs in the small intestine through muscular constriction of the intestinal wall. This process, called segmentation, is similar to peristalsis, except that the rhythmic timing of the muscle constrictions forces the food backward and forward rather than forward only.

5. *Chemical digestion* is the process of chemically breaking down food into simpler molecules. The process is carried out by enzymes in the stomach and small intestines.

6. *Absorption* is the movement of molecules (by passive diffusion or active transport) from the digestive tract to adjacent blood and lymphatic vessels. Absorption is the entrance of the digested food (now called nutrients) into the body.

7. *Defecation* is the process of eliminating undigested material through the anus.

Structure of the Digestive Tract Wall

The digestive tract, from the esophagus to the anus, is characterized by a wall with four layers, or tunics. The layers are discussed below, from the inside lining of the tract to the outside lining:

■ The mucosa is a mucous membrane that lines the inside of the digestive tract from mouth to anus. Depending on the section of the digestive tract, it protects the digestive tract wall, secretes substances, and absorbs the end products of digestion. It is composed of three layers:

  ■ The epithelium is the innermost layer of the mucosa. It is composed of simple columnar epithelium or stratified squamous epithelium. Also present are goblet cells and endocrine cells. Goblet cells secrete mucus that protects the epithelium from digestion, and endocrine cells secrete hormones into the blood.

  ■ The lamina propria lies outside the epithelium. It is composed of areolar connective tissue. Blood vessels and lymphatic vessels present in this layer provide nutrients to the epithelial layer, distribute hormones produced in the epithelium, and absorb end products of digestion from the lumen. The lamina propria also contains the **mucosa-associated lymphoid tissue (MALT)**, nodules of lymphatic tissue bearing lymphocytes and macrophages that protect the GI tract wall from bacteria and other pathogens that may be mixed with food.

  ■ The muscularis mucosae, the outer layer of the mucosa, is a thin layer of smooth muscle responsible for generating local movements. In the stomach and small intestine, the smooth muscle generates folds that increase the absorptive surface area of the mucosa.

■ The submucosa lies outside the mucosa. It consists of areolar connective tissue containing blood vessels, lymphatic vessels, and nerve fibers.

■ The muscularis (muscularis externa) is a layer of muscle. In the mouth and pharynx, it consists of skeletal muscle that aids in swallowing. In the rest of the digestive tract, it consists of smooth muscle (three layers in the stomach, two layers in the small and large intestines) and associated nerve fibers. The smooth muscle is responsible for movement of food by peristalsis and mechanical digestion by segmentation. In some regions, the circular layer of smooth muscle enlarges to form sphincters, circular muscles that control the opening and closing of the lumen (such as between the stomach and small intestine).

■ The serosa is a serous membrane that covers the muscularis externa of the digestive tract in the peritoneal cavity. The following is a description of the various types of serosae associated with the digestive system:

   ■ The adventitia is the serous membrane that lines the muscularis externa of the oral cavity, pharynx, esophagus, and rectum.

   ■ The visceral peritoneum is the serous membrane that lines the stomach, large intestine, and small intestine.

   ■ The mesentery is an extension of the visceral peritoneum that attaches the small intestine to the rear abdominal wall.

   ■ The mesocolon is an extension of the visceral peritoneum that attaches the large intestine to the rear of the abdominal wall.

   ■ The parietal peritoneum lines the abdominopelvic cavity (abdominal and pelvic cavities). The abdominal cavity contains the stomach, small intestine, large intestine, liver, spleen, and pancreas. The pelvic cavity contains the urinary bladder, rectum, and internal reproductive organs.

Digestive Enzymes

During digestion, four different groups of molecules are commonly encountered. Each is broken down into its molecular components by specific enzymes:

■ Complex carbohydrates, or polysaccharides (such as starches), are broken down into oligosaccharides (consisting of two to ten linked monosaccharides), disaccharides (such as maltose), or individual monosaccharides (such as glucose or fructose). Enzymes called amylases break down starch.

■ Proteins are broken down into short chains of amino acids (peptides) or individual amino acids by enzymes called proteases.

■ Lipids are broken down into glycerol and fatty acids by enzymes called lipases.

■ Nucleic acids are broken down into nucleotides by enzymes called nucleases.

A summary of enzymes and their substrates (substances upon which enzymes operate) appears in Table 18-1.

Table 18-1   Enzymes and Their Substrates

Enzyme	Substrate	Products of Enzyme Activity
Saliva		
salivary amylase	starches	maltose, oligosaccharides
Gastric enzyme (chief cells of stomach)		
pepsin	proteins	peptides
Pancreatic enzymes (acinar cells of pancreas)		
pancreatic amylase	starches	maltose, oligosaccharides
trypsin	proteins	peptides
chymotrypsin	proteins	peptides
carboxypeptidase	proteins	peptides, amino acids
pancreatic lipase	fats	fatty acids, monoglycerides
nucleases	RNA, DNA	nucleotides
Small intestine (brush border cells on the villi of the small intestine)		
dextrinase	oligosaccharides	glucose
maltase	maltose	glucose
sucrase	sucrose	glucose, fructose
lactase	lactose	glucose, galactose
aminopeptidase	peptides	peptides, amino acids
dipeptidase	dipeptides	amino acids
nucleosidases	nucleotides	nitrogen bases, ribose, deoxyribose, phosphates
phosphatases	nucleotides	nitrogen bases, ribose, deoxyribose, phosphates

The Mouth

The mouth (oral cavity, buccal cavity) is where food enters the digestive tract. The following features are found in the mouth:

■ The vestibule is the narrow region between the cheeks and teeth and between the lips and teeth.

■ The tongue defines the lower boundary of the mouth. It helps position the food during mastication (chewing) and gathers the chewed food into a ball, or **bolus,** in preparation for swallowing. The tongue is covered with papillae, small projections that help the tongue grip food. Many of the papillae bear taste buds.

■ The palate defines the upper boundary of the mouth. The forward portion is the hard palate, hard because bone (maxillae and palatine) makes up this portion of it. Further back in the mouth, the soft palate consists of muscle and lacks any bone support. A conical muscular projection, the uvula, is suspended from the rear of the soft palate.

■ Saliva contains water (99.5 percent), digestive enzymes, lysozyme (an enzyme that kills bacteria), proteins, antibodies (IgA), and various ions. Saliva lubricates the mouth, moistens food during chewing, protects the mouth against pathogens, and begins the chemical digestion of food. Chemical digestion is carried out by the digestive enzyme salivary amylase, which breaks down polysaccharides (starch) into short chains of glucose, especially the disaccharide maltose (which consists of two glucose molecules). Saliva is produced by the following glands:

   ■ There are three pairs of salivary glands: the parotid (located near the masseter muscle), submandibular (located deep in the mandible), and sublingual (located under the tongue). They deliver their secretions to the mouth via ducts.

   ■ Buccal glands are located in the mucosa that lines the mouth.

■ The teeth are embedded within sockets of the upper and lower jawbones (maxillae and mandible). Each tooth is surrounded by gum, or gingival, and held in its socket by a periodontal ligament. The 20 deciduous teeth (baby teeth or milk teeth) are eventually replaced by 32 permanent teeth. There are three types of teeth, based on shape and function:

   ■ Incisors have a chisel-shaped edge suitable for biting off food.

   ■ Cuspids are pointed fangs and are used for tearing food.

   ■ Premolars (bicuspids) and molars have flat surfaces used for grinding and crushing food.

A tooth has the following structural features:

- Dentin is a calcified tissue (like bone) that makes up the bulk of the tooth.

- The crown is the portion of the tooth embedded in the bone.

- The neck is the region at the gumline where the crown and root meet.

- Enamel is the hard, nonliving material that covers the crown. Calcium compounds make the enamel the hardest substance in the body.

- Cementum is the bonelike substance that covers the root and binds it to the periodontal ligament.

- The pulp cavity is the central cavity inside the tooth. It contains blood vessels, nerves, and connective tissue (collectively called pulp).

The Pharynx

The pharynx, or throat, receives the food from the mouth during swallowing. From the mouth, it moves back and down into the oropharynx and then descends into the laryngopharynx. The food then passes into the esophagus.

The Esophagus

The esophagus is a 25-cm (10-inch) long tube that begins at the laryngopharynx and descends behind the trachea through the mediastinum (cavity between the lungs). It then passes through the diaphragm at an opening called the esophageal hiatus. It then turns just a bit to the left and connects to the stomach.

Food is forced through the esophagus toward the stomach by peristalsis. Two sphincter muscles, the upper esophageal sphincter at the top of the esophagus and the lower esophageal sphincter at the bottom of the esophagus, control the movement of food into and out of the esophagus.

Deglutition (Swallowing)

Swallowing, or **deglutition,** is divided into three phases:

1. The **buccal phase** occurs voluntarily in the mouth when the tongue forces the bolus of food toward the pharynx.

2. The pharyngeal phase occurs involuntarily when food enters the pharynx, as follows:

   a. The soft palate and uvula fold upward and cover the nasopharynx to prevent the passage of food up and into the nasal cavity.

   b. The epiglottis, a flexible cartilaginous flap at the top of the larynx, folds down as the larynx rises. As a result, the opening to the trachea is covered, and food can pass only into the esophagus.

3. The esophageal phase occurs involuntarily in the esophagus. The esophageal sphincter, normally closed, opens to allow food to pass when the larynx rises during swallowing. When food reaches the lower end of the esophagus, the lower esophageal sphincter opens to allow the food to enter the stomach.

The Stomach

The stomach is a J-shaped, baglike organ that expands to store food (Figure 18-1). Typical of that of the entire digestive tract, the wall of the stomach contains four layers. However, the inner layer, the mucosa, is modified for the specialized functions of the stomach. In particular, the innermost layer of the mucosa (facing the lumen) contains a layer of simple columnar epithelium consisting of goblet cells. Gastric pits on the surface penetrate deep into the layer, forming ducts whose walls are lined with various gastric glands. A summary of the glands in the mucosa follows:

■ Mucous surface cells are the goblet cells that make up the surface layer of the simple columnar epithelium. These cells secrete mucus, which protects the mucosa from the action of acid and digestive enzymes.

■ Parietal (oxyntic) cells are scattered along the neck and lower walls of the ducts. They secrete hydrochloric acid (HC) and intrinsic factor. Intrinsic factor is necessary for the absorption of vitamin B_{12} in the small intestine.

■ Chief (zymogenic) cells also line the lower walls of the ducts. They secrete pepsinogen, the inactive form of pepsin. Pepsin is a protease, an enzyme that breaks down proteins.

■ Enteroendocrine cells secrete various hormones that diffuse into nearby blood vessels. One important hormone, gastrin, stimulates other glands in the stomach to increase their output.

Figure 18-1 The parts of the digestive system.

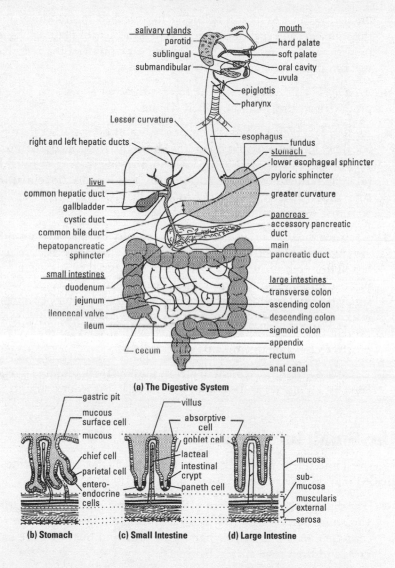

(a) The Digestive System

(b) Stomach   (c) Small Intestine   (d) Large Intestine

The first three glands listed in the preceding list are exocrine glands, whose secretions, collectively called gastric juice, enter the stomach and mix with food. The last gland is an endocrine gland, whose hormone secretions enter the blood supply.

The stomach serves a variety of functions:

- *Storage.* Because of its accordionlike folds (called rugae), the wall of the stomach can expand to store two to four liters of material. Temporary storage is important because you eat considerably faster than you can digest food and absorb its nutrients.

- *Mixing.* The stomach mixes the food with water and gastric juice to produce a creamy medium called **chyme.**

- *Physical breakdown.* Three layers of smooth muscles (rather than the usual two) in the muscularis externa churn the contents of the stomach, physically breaking food down into smaller particles. In addition, HCl denatures (or unfolds) proteins and loosens the cementing substances between cells (of the food). The HCl also kills most bacteria that may accompany the food.

- *Chemical breakdown.* Proteins are chemically broken down by the enzyme pepsin. Chief cells, as well as other stomach cells, are protected from self-digestion because chief cells produce and secrete an inactive form of pepsin, pepsinogen. Pepsinogen is converted to pepsin by the HCl produced by the parietal cells. Only after pepsinogen is secreted into the stomach cavity can protein digestion begin. Once protein digestion begins, the stomach is protected by the layer of mucus secreted by the mucous cells.

- *Controlled release.* Movement of chyme into the small intestine is regulated by a sphincter at the end of the stomach, the pyloric sphincter.

The Small Intestine

The small intestine (small in diameter compared to the large intestine) is divided into three sections, as shown in Figure 18-1:

- The duodenum, about 25 cm (10 inches) long, receives chyme from the stomach through the pyloric sphincter. Ducts that empty into the duodenum deliver pancreatic juice and bile from the pancreas and liver, respectively.

- The jejunum, about 2.5 m (8 feet) long, is the middle section of the small intestine.

- The ileum, about 3.6 m (12 feet) long, is the last section of the small intestine. It ends with the ileocecal valve (sphincter), which regulates the movement of chyme into the large intestine and prevents backward movement of material from the large intestine.

The functions of the small intestine include the following:

- *Mechanical digestion.* Segmentation mixes the chyme with enzymes from the small intestine and pancreas. Bile from the liver separates fat into smaller fat globules. Peristalsis moves the chyme through the small intestine.

- *Chemical digestion.* Enzymes from the small intestine and pancreas break down all four groups of molecules found in food (polysaccharides, proteins, fats, and nucleic acids) into their component molecules.

- *Absorption.* The small intestine is the primary location in the GI tract for absorption of nutrients.

  The components of carbohydrates, proteins, nucleic acids, and water-soluble vitamins are absorbed by facilitated diffusion or active transport. They are then passed to blood capillaries.

  - Vitamin B_{12}: Vitamin B_{12} combines with intrinsic factor (produced in the stomach) and is absorbed by receptor-mediated endocytosis. It is then passed to the blood capillaries.

  - Lipids and fat-soluble vitamins: Because fat-soluble vitamins and the components of lipids are insoluble in water, they are packaged and delivered to cells within water-soluble clusters of bile salts called micelles. They are then absorbed by simple diffusion and, once inside the cells, mix with cholesterol and protein to form **chylomicrons.** The chylomicrons are then passed to the lymphatic capillaries. When the lymph eventually empties into the blood, the chylomicrons are broken down by lipoprotein lipase, and the breakdown products, fatty acids and glycerol, pass through blood capillary walls to be absorbed by various cells.

  - Water and electrolytes: About 90 percent of the water in chyme is absorbed, as well as various electrolytes (ions), including Na^+, K^+, Cl^-, nitrates, calcium, and iron.

Modifications of the mucosa for its various specialized functions in the small intestine include the following:

■ The plicae circulares (circular folds) are permanent ridges in the mucosa that encircle the inside of the small intestine. The ridges force the food to spiral forward. The spiral motion helps mix the chyme with the digestive juices.

■ Villi (singular, villus) are fingerlike projections that cover the surface of the mucosa, giving it a velvety appearance. They increase the surface area over which absorption and digestion occur. The spaces between adjacent villi lead to deep cavities at the bases of the villi called intestinal crypts (**crypts of Lieberkühn**). Glands that empty into the cavities are called intestinal glands, and the secretions are collectively called intestinal juice.

■ Microvilli are microscopic extensions of the outer surface of the absorptive cells that line each villus. Because of their brushlike appearance (microscopically), the microvilli facing the lumen form the brush border of the small intestine. Like the villi; the microvilli increase the surface area over which digestion and absorption take place.

The villi of the mucosa have the following characteristics:

■ An outer epithelial layer (facing the lumen) consists of the following cell types:

    ■ Absorptive cells, the primary cell type of the epithelial layer, synthesize digestive enzymes called brush border enzymes that become embedded in the plasma membranes around the microvilli. Various nutrients in the chyme that move over the microvilli are broken down by these brush border enzymes and subsequently absorbed. Refer to Table 18-1, earlier in this chapter, for a list of these enzymes.

    ■ Goblet cells, located throughout the epithelial layer, secrete mucus that helps protect the epithelial layer from digestion.

    ■ **Enteroendocrine cells** secrete hormones into blood vessels that penetrate each villus.

    ■ Paneth cells, located in the epithelial layer facing the intestinal crypts, secrete lysozyme, an enzyme that destroys bacteria.

■ An inner core of lamina propria (connective tissues) contains blood capillaries and small lymphatic capillaries called lacteals.

The submucosa that underlies the mucosa of the small intestine bears the following modifications:

- **Brunner's (duodenal) glands,** found only in the submucosa of the duodenum, secrete an alkaline mucus that neutralizes the gastric acid in the incoming chyme.

- **Peyer's patches** (aggregated lymphatic nodules), found mostly in the submucosa of the ileum, are clusters of lymphatic nodules that provide a defensive barrier against bacteria.

Large Intestine

The large intestine is about 1.5 m (5 feet) long and is characterized by the following components:

- The cecum is a dead-end pouch at the beginning of the large intestine, just below the ileocecal valve.

- The appendix (vermiform appendix) is an 8-cm (3-inch) long, finger-like attachment to the cecum that contains lymphoid tissue and serves immunity functions.

- The colon, representing the greater part of the large intestine, consists of four sections: the ascending, transverse, descending, and sigmoid colons. At regular distances along the colon, the smooth muscle of the muscularis layer causes the intestinal wall to gather, producing a series of pouches called haustra. The epithelium facing the lumen of the colon is covered with openings of tubular intestinal glands that penetrate deep into the thick mucosa. The glands consist of absorptive cells and goblet cells. The absorptive cells absorb water and the goblet cells secrete mucus. The mucus lubricates the walls of the large intestine to smooth the passage of feces.

- The rectum is the last 20 cm (8 inches) of the large intestine. The mucosa in the rectum forms longitudinal folds called anal columns.

- The anal canal, the last 3 cm (1 inch) of the rectum, opens to the exterior at the anus. An involuntary (smooth) muscle, the interior anal sphincter, and a voluntary (skeletal) muscle, the external anal sphincter, control the release of the feces through the anus.

The functions of the large intestine include the following:

■ *Peristalsis:* Rhythmic contractions of the large intestine produce a form of segmentation called haustral contractions in which food residues are mixed and forced to move from one haustrum to the next. Peristaltic contractions produce mass movements of larger amounts of material.

■ *Bacterial digestion:* Bacteria that colonize the large intestine digest waste products. They break down indigestible material by fermentation, releasing various gases. Vitamin K and certain B vitamins are also produced by bacterial activity.

■ *Absorption:* Vitamins B and K, some electrolytes (Na$^+$ and Cl$^-$), and most of the remaining water is absorbed by the large intestine.

■ *Defecation:* Mass movement of feces into the rectum stimulates a defecation reflex that opens the internal anal sphincter. Unless the external sphincter is voluntarily closed, feces will be evacuated through the anus.

The Pancreas

The secretions of the pancreas, called pancreatic juice, include various enzymes, including pancreatic amylase (digestion of starch), trypsin, carboxypepiydase, and chymotrypsin (proteases), as well as pancreatic lipase (digestion of fats). Sodium bicarbonate is also produced, making the pancreatic juice alkaline. This alkaline solution stabilizes the pH in the duodenum, thus providing an optimal environment for the action of these enzymes.

Pancreatic juice is produced in clusters of exocrine cells called **acini.** The remaining cells in the pancreas (about 1 percent of the total) also form clusters (pancreatic islets). These are the endocrine cells that produce the hormones insulin, glucagon, somatostatin, and pancreatic polypeptide.

Pancreatic juice collects in small ducts that merge to form two large ducts. The main pancreatic duct exits the pancreas and merges with the common bile duct from the liver and gallbladder. This combined duct, called the hepatopancreatic ampulla, then enters the duodenum by passing through the hepatopancreatic sphincter. A smaller, second duct that exits the pancreas, the accessory pancreatic duct, joins the duodenum directly.

The Liver and Gallbladder

The digestive function of the liver is to produce **bile,** which is then delivered to the duodenum to emulsify fats. Emulsification is the breaking up of fat globules into smaller fat droplets, increasing the surface area upon which fat-digesting enzymes (lipases) can operate. Because bile is not involved in breaking any chemical bonds, it is not an enzyme. It is an emulsifier. Bile is also alkaline, serving to help neutralize the HCl in the chyme.

Bile consists of bile salts, bile pigments, phospholipids (including lecithin), cholesterol, and various ions. The primary bile pigment, **bilirubin,** is an end product of the breakdown of hemoglobin from expended red blood cells. The bile that is lost via the feces consists of bilirubin. This is the body's natural way of getting rid of bilirubin. Bilirubin gives the feces a brown color.

The liver performs numerous metabolic functions. Some of the most important follow:

- Bile is produced.

- Blood glucose is regulated. When blood glucose is high, the liver converts glucose to glycogen (**glycogenesis**) and stores the glycogen. When blood glucose is low, glycogen is broken down (**glycogenolysis**), and glucose is released into the blood.

- Proteins (including plasma proteins) and certain amino acids are synthesized.

- Ammonia (which is toxic) is converted to urea (less toxic) for elimination by the kidneys.

- Bacteria and expended red and white blood cells are broken down. From the red blood cells, iron and globin are recycled, and bilirubin is secreted in the bile.

- Vitamins (A, D, and B_{12}) and minerals (including iron from expended red blood cells) are stored.

- Toxic substances (drugs, poisons) and hormones are broken down.

The liver is composed of numerous functional units called lobules. Within each lobule, epithelial cells called hepatocytes are arranged in layers that radiate out from a central vein. Hepatic sinusoids are spaces that lie between groups of layers, while smaller channels called bile canaliculi separate other layers. Each of (usually) six corners of the lobule are occupied by three vessels: one bile duct and two blood vessels (a portal triad). The blood vessels are branches from the hepatic artery (carrying oxygenated blood) and from the hepatic portal vein (carrying deoxygenated but nutrient-rich blood from the small intestine).

Blood enters the liver through the hepatic artery and hepatic portal vein and is distributed to lobules. Blood flows into each lobule by passing through the hepatic sinusoid and collecting in the central vein. The central veins of all the lobules merge and exit the liver through the hepatic vein (not the hepatic portal vein).

Within the sinusoids, phagocytes called **Kupffer cells** (stellate reticuloendothelial cells) destroy bacteria and break down expended red and white blood cells and other debris. Hepatocytes that border the sinusoids also screen the incoming blood. They remove various substances from the blood, including oxygen, nutrients, toxins, and waste materials. From these substances they produce bile, which they secrete into the bile canaliculi, which empty into bile ducts. Bile ducts from the various lobules merge and exit the liver as a single common hepatic duct. The common hepatic duct merges with the cystic duct from the gallbladder to form the common bile duct, which in turn merges with the pancreatic duct to form the hepatopancreatic ampulla. This last duct delivers the bile to the duodenum.

The gallbladder stores excess bile. When food is in the duodenum, bile flows readily from the liver and gallbladder into the duodenum. When the duodenum is empty, a sphincter muscle (hepatopancreatic sphincter) closes the hepatopancreatic ampulla, and bile backs up and fills the gallbladder (refer to Figure 18-1).

Regulation of Digestion

The activities of the digestive system are regulated by both hormones and neural reflexes. Four important hormones and their effects on target cells follow:

- Gastrin is produced by enteroendocrine cells of the stomach mucosa. Effects include:

  - Stimulation of gastric juice (especially HCl) secretion by gastric glands.

  - Stimulation of smooth muscle contraction in the stomach, small intestine, and large intestine, which increases gastric and intestinal motility.

  - Relaxation of the pyloric sphincter, which promotes gastric emptying into the small intestine.

- Secretin is produced by the enteroendocrine cells of the duodenal mucosa. Effects include:

  - Stimulation of bicarbonate secretion by the pancreas, which stabilizes the pH of the chyme when released into the duodenum.

  - Stimulation of bile production by the liver.

  - Inhibition of gastric juice secretions and gastric motility, which in turn slows digestion in the stomach and retards gastric emptying.

- Cholecystokinin (CCK) is produced by the enteroendocrine cells of the duodenal mucosa. Effects include:

  - Stimulation of bile release by the gallbladder.

  - Stimulation of pancreatic juice secretion.

  - Relaxation of the hepatopancreatic ampulla and opening of the hepatopancreatic sphincter, which allows the flow of bile and pancreatic juices into the duodenum.

- Glucose insulinotropic peptide (GIP) is produced and released by the enteroendocrine cells of the duodenal mucosa in response to the presence of the glucose in the small intestine. This hormone stimulates the pancreas to begin releasing insulin. Some researchers refer to this hormone as *glucose-dependent insulinotropic peptide* (still maintaining the abbreviation of GIP; some also use GDIP).

The second regulatory agent of the digestive system is the nervous system. Stimuli that influence digestive activities may originate in the head, the stomach, or the small intestine. Based on these sites, there are three phases of digestive regulation:

1. The *cephalic phase* comprises those stimuli that originate from the head: sight, smell, taste, or thoughts of food, as well as emotional states. In response, the following reflexes are initiated:

   a. *Neural response:* Stimuli that arouse digestion are relayed to the hypothalamus, which in turn initiates nerve impulses in the parasympathetic vagus nerve. These impulses innervate nerve networks of the GI tract (enteric nervous system), which promote contraction of smooth muscle (which causes peristalsis) and secretion of gastric juice. Stimuli that repress digestion (emotions of fear or anxiety, for example) innervate sympathetic fibers that suppress muscle contraction and secretion.

   b. *General effects:* The stomach prepares for the digestion of proteins.

2. The *gastric phase* describes those stimuli that originate from the stomach. These stimuli include distention of the stomach (which activates stretch receptors), low acidity (high pH), and the presence of peptides. In response, the following reflexes are initiated:

   a. *Neural response:* Gastric juice secretion and smooth muscle contraction are promoted.

   b. *Hormonal response:* Gastrin production is promoted.

   c. *General effects:* The stomach and small intestine prepare for the digestion of chyme, and gastric emptying is promoted.

3. The *intestinal phase* describes stimuli originating in the small intestine. These include distention of the duodenum, high acidity (low pH), and the presence of chyme (especially fatty acids and carbohydrates). In response, the following reflexes are initiated:

   a. *Neural response:* Gastric secretion and gastric motility are inhibited (enterogastric reflex). Intestinal secretions, smooth muscle contraction, and bile and pancreatic juice production are promoted.

   b. *Hormonal response:* Production of secretin, CCK, and GIP is promoted.

   c. *General effects:* Stomach emptying is retarded to allow adequate time for digestion (especially fats) in the small intestine. Intestinal digestion and motility are promoted.

Chapter Check-Out

Q&A

1. Which of the following is the correct order of the processes involved in the treatment of food in the digestive tract?

    a. Ingestion, propulsion, defecation, chemical digestion, absorption, secretion, mechanical digestion

    b. Ingestion, propulsion, secretion, mechanical digestion, defecation, chemical digestion, absorption

    c. Ingestion, propulsion, secretion, mechanical digestion, chemical digestion, absorption, defecation

    d. Ingestion, secretion, propulsion, mechanical digestion, chemical digestion, absorption, defecation

2. True or False: The digestive tract, from the esophagus to the anus, is characterized by a wall with five layers, or tunics.

3. During digestion, four different groups of molecules are commonly encountered. These are complex carbohydrates, or polysaccharides, _____, fats, and _____ _____.

4. Which of the following is not part of the digestive system?

    a. Esophagus

    b. Stomach

    c. Large intestine

    d. Appendix

5. Four important hormones that help regulate digestion are gastrin, _____, cholecystokinin, and _____ _____ _____.

Answers: 1. c, **2.** F, **3.** proteins, nucleic acids, **4.** d, **5.** secretin, gastric inhibitory peptide (GIP)

Chapter 19

THE URINARY SYSTEM

Chapter Check-In

❑ Naming the basic function of the urinary system

❑ Understanding the parts of the urinary system and their basic functions

❑ Knowing the three processes of the nephron

❑ Describing the countercurrent multiplier system

Understanding how the urinary system helps maintain homeostasis by removing harmful substances from the blood and regulating water balance in the body is an important part of physiology. Your kidneys, which are the main part of the urinary system, are made up of millions of nephrons that act as individual filtering units and are complex structures themselves. The ureters, urethra, and urinary bladder complete this intricate system.

The urinary system helps maintain homeostasis by regulating water balance and by removing harmful substances from the blood. The blood is filtered by two kidneys, which produce urine, a fluid containing toxic substances and waste products. From each kidney, the urine flows through a tube, the ureter, to the urinary bladder, where it is stored until it is expelled from the body through another tube, the urethra.

Anatomy of the Kidneys

The kidneys are surrounded by three layers of tissue:

■ The renal fascia is a thin, outer layer of fibrous connective tissue that surrounds each kidney (and the attached adrenal gland) and fastens it to surrounding structures.

■ The adipose capsule is a middle layer of adipose (fat) tissue that cushions the kidneys.

■ The renal capsule is an inner fibrous membrane that prevents the entrance of infections.

Inside the kidney, three major regions are distinguished, as shown in Figure 19-1:

■ The renal cortex borders the convex side.

■ The renal medulla lies adjacent to the renal cortex. It consists of striated, cone-shaped regions called renal pyramids (medullary pyramids), whose peaks, called renal papillae, face inward. The unstriated regions between the renal pyramids are called renal columns.

■ The renal sinus is a cavity that lies adjacent to the renal medulla. The other side of the renal sinus, bordering the concave surface of the kidney, opens to the outside through the renal hilus. The ureter, nerves, and blood and lymphatic vessels enter the kidney on the concave surface through the renal hilus. The renal sinus houses the renal pelvis, a funnel-shaped structure that merges with the ureter.

Blood and nerve supply

Because the major function of the kidneys is to filter the blood, a rich blood supply is delivered by the large renal arteries. The renal artery for each kidney enters the renal hilus and successively branches into segmental arteries and then into interlobar arteries, which pass between the renal pyramids toward the renal cortex. The interlobar arteries then branch into the arcuate arteries, which curve as they pass along the junction of the renal medulla and cortex. Branches of the arcuate arteries, called interlobular arteries, penetrate the renal cortex, where they again branch into afferent arterioles, which enter the filtering mechanisms, or glomeruli, of the nephrons.

Figure 19-1 (a) The urinary system, (b) the kidney, (c) cortical nephron, and (d) juxtamedullary nephron of the kidneys.

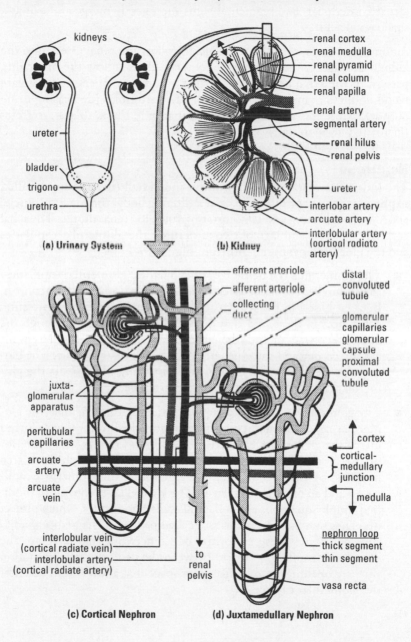

(a) Urinary System

- kidneys
- ureter
- bladder
- trigone
- urethra

(b) Kidney

- renal cortex
- renal medulla
- renal pyramid
- renal column
- renal papilla
- renal artery
- segmental artery
- renal hilus
- renal pelvis
- ureter
- interlobar artery
- arcuate artery
- interlobular artery (cortical radiate artery)

(c) Cortical Nephron

- juxta-glomerular apparatus
- peritubular capillaries
- arcuate artery
- arcuate vein
- interlobular vein (cortical radiate vein)
- interlobular artery (cortical radiate artery)
- to renal pelvis

(d) Juxtamedullary Nephron

- efferent arteriole
- afferent arteriole
- collecting duct
- distal convoluted tubule
- glomerular capillaries
- glomerular capsule
- proximal convoluted tubule
- cortex
- cortical-medullary junction
- medulla
- nephron loop
- thick segment
- thin segment
- vasa recta

Blood leaving the nephrons exits the kidney through veins that trace the same path, in reverse, as the arteries that delivered the blood. Interlobular, arcuate, interlobar, and segmental veins successively merge and exit as a single renal vein.

Autonomic nerves from the renal plexus follow the renal artery into the kidney through the renal hilus. The nerve fibers follow the branching pattern of the renal artery and serve as vasomotor fibers that regulate blood volume. Sympathetic fibers constrict arterioles (decreasing urine output), while less numerous parasympathetic fibers dilate arterioles (increasing urine output).

Nephrons

The kidney consists of over a million individual filtering units called **nephrons**. Each nephron consists of a filtering body, the renal corpuscle, and a urine-collecting and concentrating tube, the renal tubule. The renal corpuscle is an assemblage of two structures, the glomerular capillaries and the glomerular capsule, shown in Figure 19-1.

■ The glomerulus is a dense ball of capillaries (glomerular capillaries) that branches from the afferent arteriole that enters the nephron. Because blood in the glomerular capillaries is under high pressure, substances in the blood that are small enough to pass through the pores (fenestrae, or endothelial fenestrations) in the capillary walls are forced out and into the encircling glomerular capsule. The glomerular capillaries merge, and the remaining blood exits the glomerular capsule through the efferent arteriole.

■ The glomerular capsule is a cup-shaped body that encircles the glomerular capillaries and collects the material (filtrate) that is forced out of the glomerular capillaries. The filtrate collects in the interior of the glomerular capsule, the capsular space, which is an area bounded by an inner visceral layer (that faces the glomerular capillaries) and an outer parietal layer. The visceral layer consists of modified simple squamous epithelial cells called podocytes, which project branches that bear fine processes called pedicels. The pedicels' adjacent podocytes mesh to form a dense network that envelops the glomerular capillaries. Spaces between the pedicels, called filtration slits, are openings into the capsular space that allow filtrate to enter the glomerular capsule.

- The renal tubule consists of three sections:

  - The first section, the proximal convoluted tubule (PCT), exits the glomerular capsule as a winding tube in the renal cortex. The wall of the PCT consists of cuboidal cells containing numerous mitochondria and bearing a brush border of dense microvilli that face the lumen (interior cavity). The high-energy yield and large surface area of these cells support their functions of reabsorption and secretion.

  - The middle of the tubule, the nephron loop, is shaped like a hairpin and consists of a descending limb that drops into the renal medulla and an ascending limb that rises back into the renal cortex. As the loop descends, the tubule suddenly narrows, forming the thin segment of the loop. The loop subsequently widens in the ascending limb, forming the thick segment of the loop. Cells of the nephron loop vary from simple squamous epithelium (descending limb and thin segment of ascending limb) to cuboidal and low columnar epithelium (thick segment of ascending limb) and almost entirely lack microvilli.

  - The final section, the distal convoluted tubule (DCT), coils within the renal cortex and empties into the collecting duct. Cells here are cuboidal with few microvilli.

Renal tubules of neighboring nephrons empty urine into a single collecting duct. Here and in the final portions of the DCT, there are cells that respond to the hormones aldosterone and antidiuretic hormone (ADH), and there are cells that secrete H+ in an effort to maintain proper pH.

Various collecting ducts within the medullary pyramids merge to form papillary ducts, which drain eventually into the renal pelvis through the medullary papillae. Urine collects in the renal pelvis and drains out of the kidney through the ureter.

The efferent arteriole carries blood away from the glomerular capillaries to form peritubular capillaries. These capillaries weave around the portions of the renal tubule that lie in the renal cortex. In portions of the nephron loop that descend deep into the renal medulla, the capillaries form loops, called vasa recta, that cross between the ascending and descending limbs. The peritubular capillaries collect water and nutrients from the filtrate in the tubule. They also release substances that are secreted into the tubule to combine with the filtrate in the formation of urine. The capillaries ultimately merge into an interlobular vein, which transports blood away from the nephron region.

There are two kinds of nephrons:

- Cortical nephrons, representing 85 percent of the nephrons in the kidney, have nephron loops that descend only slightly into the renal medulla (refer to Figure 19-1).

- Juxtamedullary nephrons have long nephron loops that descend deep into the renal medulla. Only juxtamedullary nephrons have vasa recta that traverse their nephron loops (refer to Figure 19-1).

The **juxtaglomerular apparatus (JGA)** is an area of the nephron where the afferent arteriole and the initial portion of the distal convoluted tubule are in close contact. Here, specialized smooth muscle cells of the afferent arteriole, called granular juxtaglomerular (JG) cells, act as mechanoreceptors that monitor blood pressure in the afferent arteriole. In the adjacent distal convoluted tubule, specialized cells, called macula densa, act as chemoreceptors that monitor the concentration of Na^+ and Cl^- in the urine inside the tubule. Together, these cells help regulate blood pressure and the production of urine in the nephron.

The operation of the human nephron consists of three processes:

- Glomerular filtration
- Tubular reabsorption
- Tubular secretion

These three processes, which determine the quantity and quality of the urine, are discussed in the following sections.

Glomerular filtration

When blood enters the glomerular capillaries, water and solutes are forced into the glomerular capsule. Passage of cells and certain molecules are restricted as follows:

- The **fenestrae** (pores) of the capillary endothelium are large, permitting all components of blood plasma to pass except blood cells.

- A basement membrane (consisting of extracellular material) that lies between the capillary endothelium and the visceral layer of the glomerular capsule blocks the entrance of large proteins into the glomerular capsule.

- The filtration slits between the pedicels of the podocytes prevent the passage of medium-sized proteins into the glomerular capsule.

The **net filtration pressure (NFP)** determines the quantity of filtrate that is forced into the glomerular capsule. The NFP, estimated at about 10 mm Hg, is the sum of pressures that promote filtration less the sum of those that oppose filtration. The following contribute to the NFP:

- The **glomerular hydrostatic pressure** (blood pressure in the glomerulus) promotes filtration.

- The **glomerular osmotic pressure** inhibits filtration. This pressure is created as a result of the movement of water and solutes out of the glomerular capillaries, while proteins and blood cells remain. This increases the concentration of solutes (thus decreasing the concentration of water) in the glomerular capillaries and therefore promotes the return of water to the glomerular capillaries by osmosis.

- The **capsular hydrostatic pressure** inhibits filtration. This pressure develops as water collects in the glomerular capsule. The more water in the capsule, the greater the pressure.

The **glomerular filtration rate (GFR)** is the rate at which filtrate collectively accumulates in the glomerulus of each nephron. The GFR, about 125 mL/min (180 liters/day), is regulated by the following:

- Renal autoregulation is the ability of the kidney to maintain a constant GFR even when the body's blood pressure fluctuates. Autoregulation is accomplished by cells in the juxtaglomerular apparatus that decrease or increase secretion of a vasoconstrictor substance that dilates or constricts, respectively, the afferent arteriole.

- Neural regulation of GFR occurs when vasoconstrictor fibers of the sympathetic nervous system constrict afferent arterioles. Such stimulation may occur during exercise, stress, or other fight-or-flight conditions and results in a decrease in urine production.

■ Hormonal control of GFR is accomplished by the renin/ angiotensinogen mechanism. When cells of the juxtaglomerular apparatus detect a decrease in blood pressure in the afferent arteriole or a decrease in solute (Na^+ and Cl^-) concentrations in the distal tubule, they secrete the enzyme renin. Renin converts angiotensinogen (a plasma protein produced by the liver) to angiotensin I. Angiotensin I in turn is converted to angiotensin II by the angiotensin-converting enzyme (ACE), an enzyme produced principally by capillary endothelium in the lungs. Angiotensin II circulates in the blood and increases GFR by doing the following:

■ Constricting blood vessels throughout the body, causing the blood pressure to rise

■ Stimulating the adrenal cortex to secrete aldosterone, a hormone that increases blood pressure by decreasing water output by the kidneys

Tubular reabsorption

In healthy kidneys, nearly all of the desirable organic substances (proteins, amino acids, glucose) are reabsorbed by the cells that line the renal tube. These substances then move into the peritubular capillaries that surround the tubule. Most of the water (usually more than 99 percent of it) and many ions are reabsorbed as well, but the amounts are regulated so that blood volume, pressure, and ion concentration are maintained within required levels for homeostasis.

Reabsorbed substances move from the lumen of the renal tubule to the lumen of a peritubular capillary. Three membranes are traversed:

■ The luminal membrane, or the side of the tubule cells facing the tubule lumen

■ The basolateral membrane, or the side of the tubule cells facing the interstitial fluids

■ The endothelium of the capillaries

Tight junctions between tubule cells prevent substances from leaking out between the cells. Movement of substances out of the tubule, then, must occur through the cells, either by active transport (requiring ATP) or by passive transport processes. Once outside of the tubule and in the interstitial fluids, substances move into the peritubular capillaries or vasa recta by passive processes.

The reabsorption of most substances from the tubule to the interstitial fluids requires a membrane-bound transport protein that carries these substances across the tubule cell membrane by active transport. When all of the available transport proteins are being used, the rate of reabsorption reaches a transport maximum (Tm), and substances that cannot be transported are lost in the urine.

The following mechanisms direct tubular reabsorption in the indicated regions:

■ *Active transport of Na^+ (in the PCT, DCT, and collecting duct).* Because Na^+ concentration is low inside tubular cells, Na^+ enters the tubular cells (across the luminal membrane) by passive diffusion. At the other side of the tubule cells, the basolateral membrane bears proteins that function as sodium-potassium (Na^+-K^+) pumps. These pumps use ATP to simultaneously export Na^+ while importing K^+. Thus, Na^+ in the tubule cells is transported out of the cells and into the interstitial fluid by active transport. The Na^+ in the interstitial fluid then enters the capillaries by passive diffusion. (The K^+ that is transported into the cell leaks back passively into the interstitial fluid.)

■ *Symporter transport (secondary active transport) of nutrients and ions (in the PCT and nephron loop).* Various nutrients, such as glucose and amino acids, and certain ions (K^+ and Cl^-) in the thick ascending limb of the nephron loop are transported into the tubule cells by the action of Na^+ symporters. A Na^+ symporter is a transport protein that carries both Na^+ and another molecule, such as glucose, across a membrane in the same direction. Movement of glucose and other nutrients from the tubular lumen into the tubule cells occurs in this fashion. The process requires a low concentration of Na^+ inside the cells, a condition maintained by the Na^+-K^+ pump operating on the basolateral membranes of the tubule cells. The movement of nutrients into cells by this mechanism is referred to as secondary active transport, because the ATP-requiring mechanism is the Na^+-K^+ pump and not the symporter itself. Once inside the tubular cells, nutrients move into the interstitial fluid and into the capillaries by passive processes.

■ *Passive transport of H_2O by osmosis (in the PCT and DCT).* The buildup of Na^+ in the peritubular capillaries creates a concentration gradient across which water passively moves, from tubule to capillaries, by osmosis. Thus, the reabsorption of Na^+ by active transport generates the subsequent reabsorption of H_2O by passive transport, a process called obligatory H_2O reabsorption.

■ *Passive transport of various solutes by diffusion (in the PCT and DCT, and collecting duct).* As H_2O moves from the tubule to the capillaries, various solutes such as K^+, Cl^-, HCO_3^-, and urea become more concentrated in the tubule. As a result, these solutes follow the water, moving by diffusion out of the tubule and into capillaries where their concentrations are lower, a process called solvent drag. Also, the accumulation of the positively charged Na^+ in the capillaries creates an electrical gradient that attracts (by diffusion) negatively charged ions (Cl^-, HCO_3^-).

■ *H_2O and solute transport regulated by hormones (in the DCT and collecting duct).* The permeability of the DCT and collecting duct and the resultant reabsorption of H_2O and Na^+ are controlled by two hormones:

   ■ Aldosterone increases the reabsorption of Na^+ and H_2O by stimulating an increase in the number of Na^+-K^+ pump proteins in the principal cells that line the DCT and collecting duct.

   ■ Antidiuretic hormone (ADH) increases H_2O reabsorption by stimulating an increase in the number of H_2O-channel proteins in the principal cells of the collecting duct.

Tubular secretion

In contrast to tubular reabsorption, which returns substances to the blood, tubular secretion removes substances from the blood and secretes them into the filtrate. Secreted substances include H^+, K^+, NH_4^+ (ammonium ion), creatinine (a waste product of muscle contraction), and various other substances (including penicillin and other drugs). Secretion occurs in portions of the PCT, DCT, and collecting duct.

■ *Secretion of H^+.* Because a decrease in H^+ causes a rise in pH (a decrease in acidity), H^+ secretion into the renal tubule is a mechanism for raising blood pH. Various acids produced by cellular metabolism accumulate in the blood and require that their presence be neutralized by removing H^+. In addition, CO_2, also a metabolic byproduct, combines with water (catalyzed by the enzyme carbonic anhydrase) to produce carbonic acid (H_2CO_3), which dissociates to produce H^+, as follows:

$$CO_2 + H_2O \longleftrightarrow H_2CO_3 \longleftrightarrow H^+ + HCO_3^-$$

This chemical reaction occurs in either direction (it is reversible) depending on the concentration of the various reactants. As a result, if HCO_3^- increases in the blood, it acts as a buffer of H^+, combining with it (and effectively removing it) to produce CO_2 and H_2O. CO_2 in tubular cells of the collecting duct combines with H_2O to form H^+ and HCO_3^-. The CO_2 may originate in the tubular cells or it may enter these cells by diffusion from the renal tubule, interstitial fluids, or peritubular capillaries. In the tubule cell, Na^+/H^+ antiporters, enzymes that move transported substances in opposite directions, transport H^+ across the luminal membrane into the tubule while importing Na^+. Inside the tubule, H^+ may combine with any of several buffers that entered the tubule as filtrate (HCO_3^-, NH_3, or HPO_4^{2-}). If HCO_3^- is the buffer, then H_2CO_3 is formed, producing H_2O and CO_2. The CO_2 then enters the tubular cell, where it can combine with H_2O again. If H^+ combines with another buffer, it is excreted in the urine. Regardless of the fate of the H+ in the tubule, the HCO_3^- produced in the first step is transported across the basolateral membrane by an HCO_3^-/Cl^- antiporter. The HCO_3^- enters the peritubular capillaries, where it combines with the H^+ in the blood and increases the blood pH. Note that the blood pH is increased by adding HCO_3^- to the blood, not by removing H^+.

- *Secretion of NH_3.* When amino acids are broken down, they produce toxic NH_3. The liver converts most NH_3 to urea, a less toxic substance. Both enter the filtrate during glomerular filtration and are excreted in the urine. However, when the blood is very acidic, the tubule cells break down the amino acid glutamate, producing NH_3 and HCO_3^-. The NH_3 combines with H^+, forming NH_4^+, which is transported across the luminal membrane by a Na^+ antiporter and excreted in the urine. The HCO_3^- moves to the blood (as discussed earlier for H^+ secretion) and increases blood pH.

- *Secretion of K^+.* Nearly all of the K^+ in filtrate is reabsorbed during tubular reabsorption. When reabsorbed quantities exceed body requirements, excess K^+ is secreted back into the filtrate in the collecting duct and final regions of the DCT. Because aldosterone stimulates an increase in Na^+/K^+ pumps, K^+ secretion (as well as Na^+ reabsorption) increases with aldosterone.

Regulation of Urine Concentration

The nephron loop of juxtamedullary nephrons is the apparatus that allows the nephron to concentrate urine. The loop is a countercurrent multiplier system in which fluids move in opposite directions through side-by-side, semi-permeable tubes. Substances are transported horizontally, by passive or active mechanisms, from one tube to the other. The movement of the transported substances up and down the tubes results in a higher concentration of substances at the bottom of the tubes than at the top of the tubes. Details of the process follow and are also shown in Figure 19-2:

Figure 19-2 The loop is a countercurrent multiplier system in which fluids move in opposite directions through side-by-side, semi-permeable tubes. This process regulates the concentration of urine.

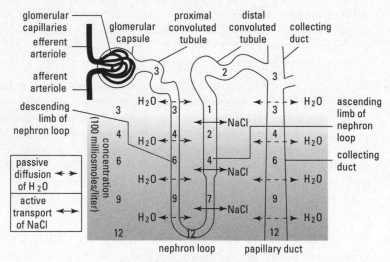

Regulation of Urine Concentration

1. The descending limb of the nephron loop is permeable to H_2O, so H_2O diffuses out into the surrounding fluids. Because the loop is impermeable to Na^+ and Cl^- and because these ions are not pumped out by active transport, Na^+ and Cl^- remain inside the loop.

2. As the fluid continues to travel down the descending limb of the loop, it becomes more and more concentrated, as water continues to diffuse out. Maximum concentration occurs at the bottom of the loop.

3. The ascending limb of the nephron loop is impermeable to water, but Na^+ and Cl^- are pumped out into the surrounding fluids by active transport.

4. As fluid travels up the ascending limb, it becomes less and less concentrated because Na^+ and Cl^- are pumped out. At the top of the ascending limb, the fluid is only slightly less concentrated than at the top of the descending limb. In other words, there is little change in the concentration of the fluid in the tubule as a result of traversing the nephron loop.

5. In the fluid surrounding the nephron loop, however, a gradient of salt (Na^+, Cl^-) is established, increasing in concentration from the top to the bottom of the loop.

■ Fluid at the top of the collecting duct has a concentration of salts about equal to that at the beginning of the nephron loop (some water is reabsorbed in the DCT). As the fluid descends the collecting duct, the fluid is exposed to the surrounding salt gradient established by the nephron loop. Without ADH, the collecting duct is impermeable to H_2O. Two outcomes are possible:

   ■ If water conservation is necessary, ADH stimulates the opening of water channels in the collecting duct, allowing H_2O to diffuse out of the duct and into the surrounding fluids. The result is concentrated urine (refer to Figure 19-2).

   ■ If water conservation is not necessary, ADH is not secreted and the duct remains impermeable to H_2O. The result is dilute urine.

■ The vasa recta delivers O_2 and nutrients to cells of the nephron loop. The vasa recta, like other capillaries, is permeable to both H_2O and salts and could disrupt the salt gradient established by the nephron loop. To avoid this, the vasa recta acts as a countercurrent multiplier system as well. As the vasa recta descends into the renal medulla, water diffuses out into the surrounding fluids, and salts diffuse in. When the vasa recta ascends, the reverse occurs. As a result, the concentration of salts in the vasa recta is always about the same as that in the surrounding fluids, and the salt gradient established by the nephron loop remains in place.

Ureters

The ureters, one from each kidney, deliver urine to the urinary bladder. The ureters enter through the back of the urinary bladder, entering at an angle such that when the urinary bladder fills, the ureter openings are forced closed. A cross section of the ureter reveals three layers of tissue:

- An inner mucosa consists of transitional epithelium covered by a lamina propria of connective tissue. Mucus secretions protect the ureter tissues from the urine.

- A middle muscularis layer consists of longitudinal and circular layers of smooth muscle fibers. The muscle fibers force urine forward by peristalsis.

- The outer adventitia consists of areolar connective tissue containing nerves, blood vessels, and lymphatic vessels.

Urinary Bladder

The urinary bladder is a muscular sac for storing urine. The triangular base of the urinary bladder, the **trigone,** is defined by the two ureters that deliver the urine and the one urethra that drains the urine. When empty, the urinary bladder collapses and develops folds (called rugae) within the urinary bladder wall. As it fills, the folds become distended and the urinary bladder becomes spherical (smooth on the inside). The wall of the urinary bladder consists of three layers similar to those of the urethra: the mucosa, the muscularis (here called the **detrusor muscle**), and the **adventitia.** Circular smooth muscle fibers around the urethra form the internal urethral sphincter.

Urethra

The urethra drains urine from the urinary bladder to an exterior opening of the body, the external urethral orifice. In females, the urethra is about 3 to 4 cm (1.5 in) long and opens to the outside of the body between the vagina and the clitoris. In males, the urethra is about 15 to 20 cm (6 to 8 in) long and passes through the prostate gland, the urogenital diaphragm, and the penis. In these regions, the urethra is called the prostatic urethra, membraneous urethra, and spongy (penile) urethra, respectively. In both males and females, a skeletal muscle, the external urethral sphincter, surrounds the urethra as it passes through the urogenital diaphragm.

Micturition, or urination, is the process of releasing urin[e] bladder into the urethra. When the urinary bladder fills to 300 mL, stretch receptors in the urinary bladder trigge[r] signal stimulates the spinal cord, which responds with [a] impulse that relaxes the internal urethral sphincter and contracts the detrusor muscle. Urine does not flow, however, until a voluntary nerve impulse relaxes the skeletal muscle of the external urethral sphincter.

The internal urethral sphincter is involuntary. The external urethral sphincter is voluntary. It is the external urethral sphincter that we learned to control when going through the potty training phase of life.

Chapter Check-Out

Q&A

1. True or False: The kidney consists of over a billion *million* individual filtering units called nephrons.

2. The urinary system helps maintain _____ by regulating water balance and removing harmful substances from the blood.

3. Which of the following is not a process of the human nephron?

   a. Tubular secretion
   b. Hemostasis
   c. Glomerular filtration
   d. Tubular reabsorption

4. True or False: The ureters, one from each kidney, deliver urine to the urinary bladder.

5. The _____ _____ is a muscular sac for storing urine.

Answers: 1. F, **2.** homeostasis, **3.** b, **4.** T, **5.** urinary bladder

Chapter 20

THE REPRODUCTIVE SYSTEM

Chapter Check-In

❑ Listing the main structures of the male and female reproductive systems

❑ Knowing the structures of a sperm cell

❑ Understanding the hormonal regulation of sperm production

❑ Describing the function of the mammary glands

❑ Comprehending the ovarian cycle and the menstrual cycle

This chapter is about the human reproductive system. A basic understanding of the reproductive system of both the male and female anatomy is critical to the understanding of reproductive systems as a whole.

What Is Reproduction?

Reproduction describes the production of eggs and sperm and the processes leading to fertilization. The reproductive system consists of the primary sex organs, or gonads (testes in males and ovaries in females), which secrete hormones and produce gametes (sperm and eggs). Accessory reproductive organs include ducts, glands, and the external genitalia.

The Male Reproductive System

The male reproductive system consists of the following structures, as shown in Figure 20-1:

Figure 20-1 View of the male reproductive system.

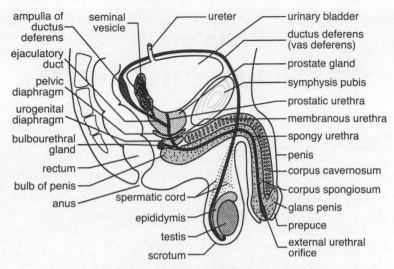

Sagittal View of Male Reproductive System

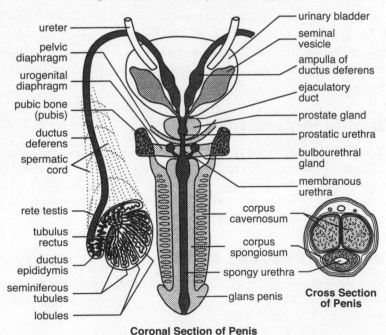

Coronal Section of Penis

■ The scrotum is a sac consisting of skin and superficial fascia that hangs from the base of the penis. A vertical septum divides the scrotum into left and right compartments, each of which encloses a testis. The external scrotum positions the testes outside the body in an environment about 3°C below that of the body cavity, a condition necessary for the development and storage of sperm. The following two muscles help maintain this temperature if the external conditions get too cold:

■ The dartos muscle is located in the superficial fascia of the scrotum and septum. Contraction of this smooth muscle creates wrinkles in the scrotum skin. The wrinkling thickens the skin, reducing heat loss when external temperatures are too cold.

■ The cremaster muscles extend from the internal oblique muscle to the scrotum. Contraction of these skeletal muscles lifts the scrotum closer to the body when external temperatures are too cold.

■ Each of the two testes (singular, testis) consists of the following structures:

■ The tunica vaginalis is a two-layer outer serous membrane surrounding each testis.

■ The tunica albuginea lies inside the tunica vaginalis and protrudes inward, dividing each testis into compartments called lobules.

■ One to four tightly coiled tubes, the seminiferous tubules, lie inside each lobule. The seminiferous tubules are the sites of sperm production (spermatogenesis). The tubule is lined with spermatogenic cells, which form sperm, and sustentacular cells (Sertoli cells), which support the developing sperm. The coiled seminiferous tubules inside each lobule unite to form a straight tube, the tubulus rectus.

■ The rete testis is a network of tubes formed by the merging of the tubules recti from each lobule.

■ The efferent ducts transport sperm out of the testis (from the rete testis) to the epididymis.

■ Interstitial cells surrounding the seminiferous tubules secrete testosterone and other androgen hormones.

■ The epididymis is a comma-shaped organ that lies adjacent to each testis. Each of the two epididymides contains a tightly coiled tube, the ductus epididymis. Here, sperm complete their maturation and are stored until ejaculation. During ejaculation, smooth muscles encircling the epididymis contract, forcing mature sperm into the next tube, the ductus deferens. The walls of the ductus epididymis contain microvilli called stereocilia that nourish sperm.

■ The ductus deferens (vas deferens) is the tube through which sperm travel when they leave the epididymis. Each of the two tubes enters the abdominal cavity, passes around the urinary bladder (refer to Figure 20-1), and together with the duct from the seminal vesicle, joins the ejaculatory duct. Before entering the ejaculatory duct, the ductus deferens enlarges, forming a region called the ampulla. Sperm are stored in the ductus deferens until peristaltic contractions of the smooth muscles surrounding the ductus force sperm forward during ejaculation.

■ The ejaculatory ducts are short tubes that connect each ductus deferens to the urethra.

■ The urethra is the passageway for urine and semen (sperm and associated secretions). Three regions of the urethra are distinguished:

■ The prostatic urethra passes through the prostate gland.

■ The membranous urethra passes through the urogenital diaphragm (muscles associated with the pelvic region).

■ The spongy (penile) urethra passes through the penis.

The urethra ends at the external urethral orifice.

■ The spermatic cord contains blood vessels, lymphatic vessels, nerves, the ductus deferens, and the cremaster muscle. It connects each testis to the body cavity, entering the abdominal wall through the inguinal canal.

■ The accessory sex glands secrete substances into the passageways that transport sperm. These substances contribute to the liquid portion of the semen:

■ The seminal vesicles secrete into the vas deferens an alkaline fluid (which neutralizes the acid in the vagina), fructose (which provides energy for the sperm), and prostaglandins (which increase sperm viability and stimulate female uterine contractions that help sperm move into the uterus).

- The prostate gland secretes a milky, slightly acidic fluid into the urethra. Various substances in the fluid increase sperm mobility and viability.

- The bulbourethral glands secrete an alkaline fluid into the spongy urethra. The fluid neutralizes acidic urine in the urethra before ejaculation occurs.

- The penis is a cylindrical organ that passes urine and delivers sperm. It consists of a root that attaches the penis to the perineum, a body (shaft) that makes up the bulk of the penis, and the glans penis, the enlarged end of the body. The glans penis is covered by a prepuce (foreskin), which may be surgically removed in a procedure called circumcision. Internally, the penis consists of three cylindrical masses of tissue, each of which is surrounded by a thin layer of fibrous tissue, the tunic albuginea. The three cylindrical masses, which function as erectile bodies, are as follows:

  - Two corpora cavernosa fill most of the volume of the penis. Their bases, called the crura (singular, crus) of the penis, attach to the urogenital diaphragm.

  - A single corpus spongiosum encloses the urethra and expands at the end to form the glans penis. The bulb of the penis, an enlargement at the base of the corpus spongiosum, attaches to the urogenital diaphragm.

During erection, parasympathetic neurons stimulate dilation of the arteries that deliver blood to the corpus cavernosa and spongiosum. As a result, blood collects in these blood vessels and causes the penis to begin to become erect. The developing erection also constricts the exiting veins of the penis. This causes even more erection. Ejaculation occurs when sympathetic neurons stimulate the discharge of sperm and supporting fluids from their various sources. During ejaculation, the sphincter muscle at the base of the urinary bladder constricts, preventing the passage of urine.

Spermatogenesis

The cells that line the walls of the seminiferous tubules are collectively called spermatogenic cells. Those cells nearest the basement membrane are called **spermatogonia.** These cells are stem cells—that is, they are capable of continuous division and remain undifferentiated, never maturing into specialized cells. Extending from the spermatogonia toward the lumen of the tubule are cells at various levels of maturity, with the most mature cells—the sperm—facing the lumen.

Spermatogenesis begins at puberty within the seminiferous tubules of the testes. The spermatogonia, each of which contains 46 chromosomes, divide by mitosis repeatedly and differentiate to produce primary spermatocytes (still diploid cells with 46 chromosomes each). The primary spermatocytes begin meiosis. During the first meiotic division (meiosis I, or the reduction division), each primary spermatocyte divides into two secondary spermatocytes, each with 23 chromosomes (haploid cells). During the second meiotic division (meiosis II, or the equatorial division), each secondary spermatocyte divides again, producing a total of four spermatids. Each spermatid still contains 23 chromosomes, but these chromosomes consist of only one chromatid (rather than the normal two chromatids).

Spermiogenesis describes the development of spermatids into mature sperm (sperm cells, or spermatozoa). At the end of this process, each sperm cell bears the following structures:

- The head of the sperm contains the haploid nucleus with 23 chromosomes. At the tip of the sperm head is the acrosome, a lysosome containing enzymes that are used to penetrate the egg. The acrosome originates from Golgi body vesicles that fuse to form a single lysosome.

- The midpiece is the first part of the tail. Mitochondria spiral around the midpiece and produce energy (ATP) used to generate the whip-like movements of the tail that propel the sperm.

- The tail is a **flagellum** consisting of the typical 9 + 2 microtubule array.

Hormonal regulation of spermatogenesis

The production of sperm is regulated by hormones, as shown in Figure 20-2:

- The hypothalamus begins secreting gonadotropin releasing hormone (GnRH) at puberty.

- GnRH stimulates the anterior pituitary to secrete follicle stimulating hormone (FSH) and luteinizing hormone (LH).

- LH stimulates the interstitial cells in the testes to produce testosterone and other male sex hormones (androgens). (In males, LH is also called interstitial cell stimulating hormone, or ICSH.)

Figure 20-2 Processes of hormone regulation in the male and female reproductive systems.

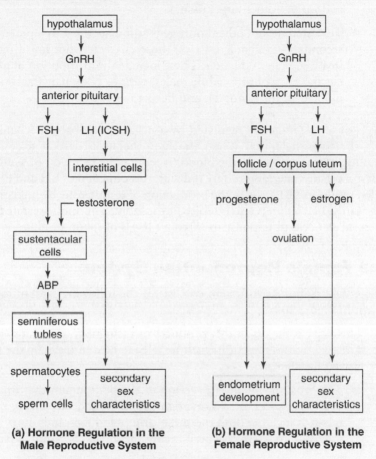

(a) Hormone Regulation in the Male Reproductive System

(b) Hormone Regulation in the Female Reproductive System

- Testosterone produces the following effects:

    - Testosterone stimulates the final stages of sperm development in the nearby seminiferous tubules. It accumulates in these tissues because testosterone and FSH act together to stimulate sustentacular cells to release androgen-binding protein (ABP). ABP holds testosterone in these cells.

- Testosterone entering the blood circulates throughout the body, where it stimulates activity in the prostate gland, seminal vesicles, and various other target tissues.

- Testosterone and other androgens stimulate the development of secondary sex characteristics, those characteristics not directly involved in reproduction. These include the distribution of muscle and fat typical in adult males, various body hair (facial and pubic hair, for example), and deepening of the voice.

Levels of testosterone are regulated by a negative-feedback mechanism with the hypothalamus. When the hypothalamus detects excessive amounts of testosterone in the blood, it reduces its secretion of GnRH. In response, the anterior pituitary reduces its production of LH and FSH, which results in a decrease in the production of testosterone by interstitial cells. GnRH secretion is also inhibited by inhibin, a hormone secreted by sustentacular cells in response to excessive levels of sperm production.

The Female Reproduction System

The female reproductive system consists of the following structures, as shown in Figure 20-3:

- The ovary is the organ that produces ova (singular, ovum), or eggs. The two ovaries present in each female are held in place by the following ligaments:

  - The broad ligament is a section of the peritoneum that drapes over the ovaries, uterus, ovarian ligament, and suspensory ligament. It includes both the mesovarium and mesometrium. The mesovarium is a fold of peritoneum that holds the ovary in place.

  - The suspensory ligament anchors the upper region of the ovary to the pelvic wall. Attached to this ligament are blood vessels and nerves, which enter the ovary at the hilus.

  - The ovarian ligament anchors the lower end of the ovary to the uterus.

  The following two tissues cover the outside of the ovary:

  - The germinal epithelium is an outer layer of simple epithelium.

  - The tunica albuginea is a fibrous layer inside the germinal epithelium.

Figure 20-3 View of the female reproductive system.

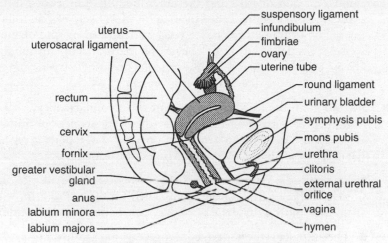

Sagittal View of Female Reproductive System

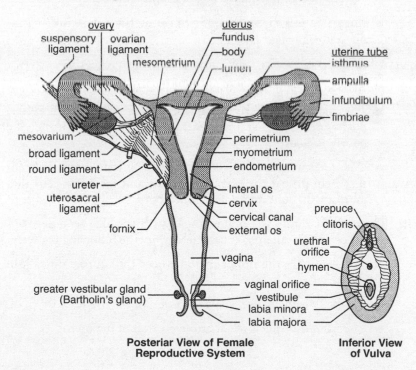

**Posterior View of Female
Reproductive System**

**Inferior View
of Vulva**

The inside of the ovary, or stroma, is divided into two indistinct regions, the outer cortex and the inner medulla. Embedded in the cortex are saclike bodies called ovarian follicles. Each ovarian follicle consists of an immature oocyte (egg) surrounded by one or more layers of cells that nourish the oocyte as it matures. The surrounding cells are called follicular cells, if they make up a single layer, or granulosa cells, if more than one layer is present.

■ The uterine tubes (oviducts) transport the secondary oocytes away from the ovary and toward the uterus (the ovaries consist of primary oocytes, which develop into secondary oocytes). The following regions characterize each of the two uterine tubes (one for each ovary):

   ■ The infundibulum is a funnel-shaped region of the uterine tube that bears fingerlike projections called fimbriae. Pulsating cilia on the fimbriae draw the secondary oocyte into the uterine tube.

   ■ The ampulla is the widest and longest region of the uterine tube. Fertilization of the oocyte by a sperm usually occurs here.

   ■ The isthmus is a narrow region of the uterine tube whose terminus enters the uterus.

   The wall of the uterine tube consists of the following three layers:

   ■ The serosa, a serous membrane, lines the outside of the uterine tube.

   ■ The middle muscularis consists of two layers of smooth muscle that generate peristaltic contractions that help propel the oocyte forward.

   ■ The inner mucosa consists of ciliated columnar epithelial cells that help propel the oocyte forward, and secretory cells that lubricate the tube and nourish the oocyte.

■ The uterus (womb) is a hollow organ within which fetal development occurs. The uterus is characterized by the following regions:

   ■ The fundus is the upper region where the uterine ducts join the uterus.

   ■ The body is the major, central portion of the uterus.

   ■ The isthmus is the lower, narrow portion of the uterus.

- The cervix is a narrow region at the bottom of the uterus that leads to the vagina. The inside of the cervix, or cervical canal, opens to the uterus above through the internal os and to the vagina below through the external os. Cervical mucus secreted by the mucosa layer of the cervical canal serves to protect against bacteria entering the uterus from the vagina. If an oocyte is available for fertilization, the mucus becomes thin and slightly alkaline. These are attributes that promote the passage of sperm. At other times, the mucus is viscous and impedes the passage of sperm.

The uterus is held in place by the following ligaments:

- Broad ligaments

- Uterosacral ligaments

- Round ligaments

- Cardinal (lateral cervical) ligaments

The wall of the uterus consists of the following three layers:

- The perimetrium is a serous membrane that lines the outside of the uterus.

- The myometrium consists of several layers of smooth muscle and imparts the bulk of the uterine wall. Contractions of these muscles during childbirth help force the fetus out of the uterus.

- The endometrium is the highly vascularized mucosa that lines the inside of the uterus. If an oocyte has been fertilized by a sperm, the zygote (the fertilized egg) implants on this tissue. The endometrium itself consists of two layers. The stratum functionalis (functional layer) is the innermost layer (facing the uterine lumen) and is shed during menstruation. The outermost stratum basalis (basal layer) is permanent and generates each new stratum functionalis.

- The vagina (birth canal) serves both as the passageway for a newborn infant and as a depository for semen during sexual intercourse. The upper region of the vagina surrounds the protruding cervix, creating a recess called the fornix. The lower region of the vagina opens to the outside at the vaginal orifice. A thin membrane called the hymen may cover the orifice. The vaginal wall consists of the following layers:

■ The outer adventitia holds the vagina in position.

■ The middle muscularis consists of two layers of smooth muscle that permit expansion of the vagina during childbirth and when the penis is inserted.

■ The inner mucosa has no glands. But bacterial action on glycogen stored in these cells produces an acid solution that lubricates the vagina and protects it against microbial infection. The acidic environment is also inhospitable to sperm. The mucosa bears transverse ridges called rugae.

■ The vulvae (pudendum) make up the external genitalia. The following structures are included:

■ The mons pubis is a region of adipose tissue above the vagina that is covered with hair.

■ The labia majora are two folds of adipose tissue that border each side of the vagina. Hair and sebaceous and sudoriferous glands are present. Developmentally, the labia majora are analogous to the male scrotum.

■ The labia minora are smaller folds of skin that lie inside the labia majora.

■ The vestibule is the recess formed by the labia minora. It encloses the vaginal orifice, the urethral opening, and ducts from the greater vestibular glands whose mucus secretions lubricate the vestibule.

■ The clitoris is a small mass of erectile and nervous tissue located above the vestibule. Extensions of the labia minora join to form the prepuce of the clitoris, a fold of skin covering the clitoris.

Mammary glands

The mammary glands are sudoriferous (sweat) glands specialized for the production of milk. The milk-producing secretory cells form walls of bulb-shaped chambers called alveoli that join together with ducts, in grapelike fashion, to form clusters called lobules. Numerous lobules assemble to form a lobe. Each breast contains a single mammary gland consisting of 15 to 20 of these lobes. Lactiferous ducts leading away from the lobes widen into lactiferous sinuses that serve as temporary reservoirs for milk. The ducts narrow again as they lead through a protruding nipple. The nipple, whose texture is made coarse by the presence of sebaceous glands, is surrounded by a ring of pigmented skin called the areola. Contraction of myoepithelial cells surrounding the alveoli force milk toward the nipples.

The breasts begin to enlarge in females at the onset of puberty. Proliferating adipose (fat) tissue expands the breast, while suspensory ligaments attached to the underlying fascia provide support. In nonpregnant females (and in males), the glands and ducts are not fully developed.

During pregnancy, estrogen and progesterone stimulate extensive development of the mammary glands and associated ducts. After childbirth, various hormones, especially prolactin from the anterior pituitary, initiate lactation, or milk production. When neurons are stimulated by the sucking of an infant, nerve impulses activate the posterior pituitary to secrete oxytocin, which in turn stimulates contraction of the myoepithelial cells surrounding the alveoli. Milk is then forced toward the nipple (the letdown reflex).

Oogenesis

Oogenesis consists of the meiotic cell divisions that lead to the production of ova (eggs) in females. The process begins during fetal development with the fetal ovary. Diploid cells called oogonia divide by mitosis and differentiate to produce primary oocytes (still diploid with 46 chromosomes). Each primary oocyte is encircled by one or more layers of cells. The oocyte and encircling cells together are called an ovarian follicle. The primary oocytes (within their follicles) begin meiosis, but division progresses only to prophase I. They remain at this stage until puberty.

The following stages in the development of an ovarian follicle are observed:

1. The primordial follicle, the initial fetal state of the follicle, encircles the oocyte with a single layer of cells, called follicular cells.

2. The primary follicle, the next stage of follicular development, possesses two or more layers of encircling cells, now called granulosa cells.

3. The secondary follicle is distinguished by the presence of the antrum, a fluid-filled, central cavity.

4. In a mature (vesicular, or Graafian) follicle, the primary oocyte has completed meiosis I. It is the stage of follicular development that precedes ejection of the oocyte from the ovary (ovulation). The following features are observed:

a. The zona pellucida, a clear layer of glycoprotein, surrounds the oocyte.

b. The corona radiata, a ring of granulosa cells, encircles the zona pellucida.

c. Several layers of cells (theca cells) surround the granulosa cells.

5. The corpus luteum is the remains of the follicle following ovulation. It remains functional, producing estrogen, progesterone, and inhibin, until it finally degenerates.

During each menstrual cycle, one primary oocyte, enclosed in its follicle, resumes meiosis I to produce two daughter cells (each haploid with 23 chromosomes). One daughter cell, the secondary oocyte, contains most of the cytoplasm, ensuring that adequate amounts of stored food, as well as mitochondria, ribosomes, and other cytoplasmic organelles, will be available for the developing embryo. The other daughter cell, a first polar body, is much smaller and contains little cytoplasm and few if any organelles. The secondary oocyte then begins meiosis II (equatorial division) but again stops at prophase (this time prophase II). The first polar body may also begin meiosis II, but it will eventually degenerate.

Ovulation occurs when a secondary oocyte and its first polar body, surrounded by the zona pellucida and corona radiata, rupture from their mature follicle and are expelled from the surface of the ovary. The oocyte is then swept up into the uterine (fallopian) tube and advances toward the uterus. If a sperm cell penetrates the corona radiata and zona pellucida and enters the secondary oocytes, meiosis II resumes in the secondary oocytes, producing an ovum and a second polar body. If a first polar body is present, it too, may resume meiosis II, producing daughter polar bodies. Fertilization occurs when the nuclei of the sperm cell and ovum unite, forming a zygote (fertilized egg). Any polar bodies present ultimately degenerate.

Hormonal regulation of oogenesis and the menstrual cycle

The human female reproductive cycle is characterized by events in the ovary (ovarian cycle) and the uterus (menstrual cycle). The purpose of these cycles is to produce an egg and to prepare the uterus for the implantation of the egg, should it become fertilized. The ovarian cycle consists of three phases:

1. The follicular phase describes the development of the follicle, the meiotic stages of division leading to the formation of the secondary oocytes, and the secretion of estrogen from the follicle.

2. Ovulation, occurring at midcycle, is the ejection of the egg from the ovary.

3. The luteal phase describes the secretion of estrogen and progesterone from the corps luteum (previously the follicle) after ovulation.

The menstrual (uterine) cycle consists of three phases:

1. The proliferative phase describes the thickening of the endometrium of the uterus, replacing tissues that were lost during the previous menstrual cycle.

2. The secretory phase follows ovulation and describes further thickening and vascularization of the endometrium in preparation for the implantation of a fertilized egg.

3. The menstrual phase (menstruation, menses) describes the shedding of the endometrium when implantation does not occur.

The activities of the ovary and the uterus are coordinated by negative- and positive-feedback responses involving gonadotropin releasing hormone (GnRH) from the hypothalamus, follicle stimulating hormone (FSH) and luteinizing hormone (LH) from the anterior pituitary, and the hormones estrogen and progesterone from the follicle and corpus luteum. A description of the events follows (see Figures 20-2 and 20-4):

1. *The hypothalamus and anterior pituitary initiate the reproductive cycle:* The hypothalamus monitors the levels of estrogen and progesterone in the blood. In a negative-feedback fashion, low levels of these hormones stimulate the hypothalamus to secrete GnRH, which in turn stimulates the anterior pituitary to secrete FSH and LH.

2. *The follicle develops:* FSH stimulates the development of the follicle from primary through mature stages.

3. *The follicle secretes estrogen:* LH stimulates the cells of the theca interna and the granulosa cells of the follicle to secrete estrogen. Inhibin is also secreted by the granulosa cells.

4. *Ovulation occurs:* Positive feedback from rising levels of estrogen stimulate the anterior pituitary (through GnRH from the hypothalamus) to produce a sudden midcycle surge of LH. This high level of LH stimulates meiosis in the primary oocyte to progress toward prophase II and triggers ovulation.

Figure 20-4 The female reproductive cycle broken down by days.

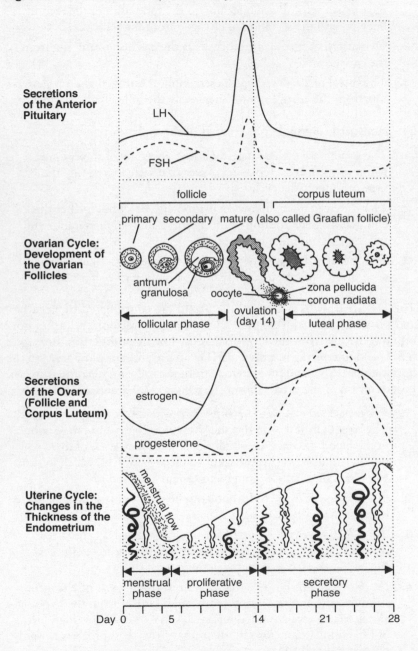

Secretions of the Anterior Pituitary

LH

FSH

follicle | corpus luteum

Ovarian Cycle: Development of the Ovarian Follicles

primary secondary mature (also called Graafian follicle)

antrum
granulosa

oocyte

zona pellucida
corona radiata

follicular phase | ovulation (day 14) | luteal phase

Secretions of the Ovary (Follicle and Corpus Luteum)

estrogen

progesterone

Uterine Cycle: Changes in the Thickness of the Endometrium

menstrual flow

menstrual phase | proliferative phase | secretory phase

Day 0   5   14   21   28

5. *The corpus luteum secretes estrogen and progesterone:* After ovulation, the follicle, now transformed into the corpus luteum, continues to develop under the influence of LH and secretes both estrogen and progesterone.

6. *The endometrium thickens:* Estrogen and progesterone stimulate the development of the endometrium, the inside lining of the uterus. It thickens with nutrient-rich tissue and blood vessels in preparation for the implantation of a fertilized egg.

7. *The hypothalamus and anterior pituitary terminate the reproductive cycle:* Negative feedback from the high levels of estrogen and progesterone cause the anterior pituitary (through the hypothalamus) to abate the production of FSH and LH. Inhibin also suppresses production of FSH and LH.

8. *The endometrium either disintegrates or is maintained, depending on whether implantation of the fertilized egg occurs, as follows:*

    Implantation does not occur: In the absence of FSH and LH, the corpus luteum deteriorates. As a result, estrogen and progesterone production stops. Without estrogen and progesterone, growth of the endometrium is no longer supported, and it disintegrates, sloughing off during menstruation.

    Implantation occurs: The implanted embryo secretes human chorionic gonadotropin (hCG) to sustain the corpus luteum. The corpus luteum continues to produce estrogen and progesterone, maintaining the endometrium. (Pregnancy tests check for the presence of hCG in the urine.)

In addition to influencing the reproductive cycle, estrogen stimulates the development of secondary sex characteristics in females. These include the distribution of adipose tissue (to the breasts, hips, and mons pubis), bone development leading to a broadening of the pelvis, changes in voice quality, and growth of various body hair.

Chapter Check-Out

Q&A

1. The development of spermatids into mature sperm (sperm cells, or spermatozoa) is called _____.

2. Which of the following is not true about testosterone?
   a. It stimulates the final stages of sperm development.
   b. It stimulates activity in the prostate gland, seminal vesicles, and various other target tissues.
   c. It neutralizes acid in the vagina.
   d. It stimulates the development of secondary sex characteristics.

3. The _____ _____ are sudoriferous (sweat) glands specialized for the production of milk.

4. True or False: The menstrual (uterine) cycle consists of three phases: the proliferative phase, the secretory phase, and the luteal phase.

5. True or False: Accessory reproductive organs include ducts, glands, and the external genitalia.

Answers: 1. spermiogenesis, **2.** c, **3.** mammary glands, **4.** F, **5.** T

REVIEW QUESTIONS

Use these review questions to practice what you've learned in this book. After you work through the review questions, you're well on your way to achieving your goal of understanding the basic concepts and strategies of anatomy and physiology.

Chapter 1

1. Which of the following is not a chemical bond?

   a. ionic

   b. oxygen

   c. covalent

   d. hydrogen

2. Which of the following is the term that best describes the left and right kidneys?

   a. They are contralateral organs.

   b. They are inferior organs.

   c. They are ipsilateral organs.

   d. They are medial organs.

3. Proteins consist of units of _____.

   a. amino acids

   b. DNA

   c. fatty acids

   d. monosaccharides

4. A _____ is the simplest kind of carbohydrate.

5. Amylase is a/an _____ that acts as a catalyst to break down starch.

Chapter 2

6. Which of the following are recognition proteins on the surface of the cell?

    a. adhesion proteins

    b. glycoproteins

    c. plasma proteins

    d. transport proteins

7. Which of the following organelles modifies the protein that is made by the ribosomes?

    a. endoplasmic reticulum

    b. Golgi apparatus

    c. lysosomes

    d. nucleus

8. If the inside of a cell has more solutes than the outside of the cell, the inside would be considered _____.

    a. hypertonic

    b. hypotonic

    c. isotonic

9. The movement of water across a biological membrane from an area of high concentration to an area of low concentration is called _____.

10. What is the name of the phase of mitosis where the paired chromatids line up in the center of the cell?

Chapter 3

11. Which of the following cells make up the inside or outside lining of tissue of the body?

    a. adipose cells

    b. areolar cells

    c. skeletal muscle cells

    d. squamous cells

12. Which of the following is typically the longest part of a neuron?

    a. axon
    b. dendrite
    c. soma (cell body)
    d. synapse

13. Which of the following types of muscle cells are under voluntary control?

    a. cardiac muscle
    b. skeletal muscle
    c. smooth muscle

14. What is the generic name for glands that secrete hormones directly into the bloodstream?

15. What type of connective tissue makes up the tendons and ligaments?

Chapter 4

16. Which of the following cells are involved in producing a chemical that gives the body its natural skin tone or perhaps a tan?

    a. keratinocytes
    b. Langerhans cells
    c. melanocytes
    d. Merkel cells

17. Many times, the physician speaks of fascia. Which of the following is considered to be the superficial fascia of the skin?

    a. cutaneous
    b. dermis
    c. epidermis
    d. hypodermis

18. Which of the following layers of the epidermis are constantly shed (sometimes forming dandruff)?

    a. stratum basale
    b. stratum corneum
    c. stratum granulosum
    d. stratum lucidum

19. The _____ muscles are involved in goose-bump formation.

20. The _____ glands are involved in acne formation.

Chapter 5

21. The ends of long bones, such as the humerus, are called _____.

    a. diaphysis

    b. endosteum

    c. epiphysis

    d. metaphysis

22. Which of the following processes of bones would be irregular in shape and the largest in size?

    a. condyle

    b. ramus

    c. trochanter

    d. tubercle

23. Which of the following terms refers to a hole in the bone?

    a. facet

    b. foramen

    c. fossa

    d. fovea

24. What is the name of the cells that break down bone?

25. Compact bone is made of cylindrical units called _____.

Chapter 6

26. Which of the following bones is an isolated bone not connected to another bone in the body?

    a. atlas

    b. hyoid

    c. incus

    d. patella

27. All ribs attach to which of the following vertebrae?

   a. cervical

   b. lumbar

   c. thoracic

   d. ribs attach to all of the above

28. Which of the following are lateral and just a bit anterior to the foramen lacerum?

   a. carotid foramen

   b. condylar canal

   c. foramen ovale

   d. jugular foramen

29. True or False: You can determine the sex of a skeleton by counting the number of ribs it has.

30. We have _____ carpal bones per wrist and _____ tarsal bones per ankle.

Chapter 7

31. Which of the following would be a type of ball-and-socket joint?

   a. amphiarthrosis

   b. diarthrosis

   c. synarthrosis

32. Which of the following would be a type of joint that is freely moveable?

   a. cartilaginous joint

   b. fibrous joint

   c. synovial joint

33. The teeth sit in a socket in the jaw. What type of joint would be a tooth socket joint?

   a. amphiarthrosis

   b. diarthrosis

   c. synarthrosis

34. What type of movement is permitted by a synarthrosis joint?

35. The knee joint is classified functionally as a _____ joint.

Chapter 8

36. Which of the following types of muscles allow blood vessels to dilate and constrict based on environmental conditions?

   a. cardiac muscle

   b. skeletal muscle

   c. smooth muscle

37. Upon stimulation, what ion ultimately causes the tropomyosin molecule to move in such a manner to expose the binding sites so the muscle can contract?

   a. calcium

   b. phosphorus

   c. potassium

   d. sodium

38. Which of the following would indicate that you are looking at cardiac cells when viewed under a microscope?

   a. the presence of a sarcolemma

   b. the presence of intercalated discs

   c. the presence of nuclei

   d. the presence of striations

39. The _____ _____ of the muscle cells release calcium ions, which are involved in muscle contraction.

40. During muscle metabolism, energy is made for muscle contraction. What is the name of the energy molecule that is used for contraction?

Chapter 9

41. The biceps brachii and triceps brachii are examples of _____ muscles when discussed as working opposite of each other.

   a. antagonistic

   b. prime mover

   c. synergistic

42. Which of the following is the most lateral of the hamstring group of muscles?

    a. biceps femoris

    b. gastrocnemius

    c. semitendinosus

    d. vastus lateralis

43. Which of the following is the most medial muscle of the thigh?

    a. gracilis

    b. semitendinosus

    c. soleus

    d. vastus medialis

44. Name the insertion point for the triceps brachii.

45. Name the large neck muscle that extends from the mastoid process of the skull to the manubrium of the sternum.

Chapter 10

46. Which of the following glial cells provide insulation for the neurons in the CNS?

    a. astrocytes

    b. ependymal cells

    c. microglia

    d. oligodendrocytes

47. A resting neuron has mainly _____ ions located on the outer side of the membrane.

    a. calcium

    b. chloride

    c. potassium

    d. sodium

48. When the gated channels open and ions move toward the inside of the axon, an impulse occurs. Which of the following is a term that refers to the impulse moving quickly down the axon to its destination quite a distance away?

   a. action potential

   b. graded potential

   c. refractory period

   d. repolarization

49. What is the name of the cells that provide insulation for the neurons in the PNS?

50. A chemical is released from the axon end and enters into the synapse, ultimately creating an impulse in the next neuron in sequence. Examples of these chemicals are acetylcholine and dopamine. What is a general name for all of these chemicals that are released from the axon into the synapse?

Chapter 11

51. The sympathetic and parasympathetic nerves belong to which of the following branches of the nervous system?

   a. afferent nervous system

   b. autonomic nervous system

   c. central nervous system

   d. somatic nervous system

52. Which of the following is involved in producing cerebrospinal fluid?

   a. cerebellum

   b. choroid plexus

   c. pons

   d. thalamus

53. Which of the following cranial nerves is involved in eye movement?

   a. optic

   b. trigeminal

   c. trochlear

   d. vagus

54. What is the name of the membranes that surround and protect the brain and spinal cord?

55. As which branch of the autonomic nerves would the vagus nerve (cranial nerve X) be classified since it slows down the heart rate?

Chapter 12

56. As which of the following receptors are involved in detecting body position?

a. exteroceptors

b. interoceptors

c. mechanoreceptors

d. proprioceptors

57. Which of the following receptors are involved in detecting taste?

a. chemoreceptors

b. gustatory receptors

c. interoceptors

d. olfactory receptors

58. The scala vestibule and scala tympani are tubes located in the _____.

a. cochlea of the ear

b. semicircular canals of the ear

c. posterior cavity of the eye

d. anterior cavity of the eye

59. Which photoreceptor cells of the retina are stimulated so we can see color?

60. Which ossicle is attached directly to the tympanic membrane of the ear?

Chapter 13

61. The endocrine system structures produce _____.

a. enzymes

b. hormones

c. neurotransmitters

d. all of the above are correct

62. Progesterone is a/an _____ hormone.

    a. eicosanoid

    b. polypeptide

    c. protein

    d. steroid

63. Which of the following hormones would be antagonistic to calcitonin?

    a. atrial natriuretic hormone

    b. leukotrienes

    c. parathyroid hormone

    d. progesterone

64. What are the two targets for oxytocin?

65. Name the hormone involved in the formation of red blood cells.

Chapter 14

66. Which of the following is NOT true of erythrocytes?

    a. They have a nucleus.

    b. They contain hemoglobin.

    c. They transport oxygen and carbon dioxide.

    d. They appear like little donuts when viewed under a microscope.

67. When a red blood cell ages, it is decomposed by the liver cells. Which of the following is a component of red blood cell decomposition?

    a. bilirubin

    b. leukotrienes

    c. transferrin

    d. uric acid

68. The QRS wave on an ECG represents which of the following?

    a. atrial contraction

    b. one heartbeat

    c. ventricular contraction

    d. ventricular filling

69. Can a person with type A blood donate to a person with type O blood?

70. What is the name of the hormone that targets the heart to help regulate blood pressure?

Chapter 15

71. Which of the following is NOT a function of lymph nodes?

   a. destroy bacteria

   b. filter lymph

   c. produce antibodies

   d. produce toxins

72. Where is the thymus located?

   a. between the lungs in the mediastinum area

   b. in the neck region

   c. just a bit inferior to the heart

   d. just a bit superior to the spleen

73. Lymph that is drained from the left side of the body enters into which of the following blood vessels?

   a. left subclavian artery

   b. left subclavian vein

   c. inferior vena cava

   d. superior vena cava

74. The appendix is actually part of the _lymphatic_ system.

75. What is the name of the large lymphatic nodules located at the back of the throat?

Chapter 16

76. Where is the major histocompatibility complex located in reference to cells?

   a. on the membrane

   b. within the nucleus of the cell

   c. within the cytoplasm of the cell

   d. within the DNA molecule of the cell

77. B cells (B lymphocytes) produce which of the following?

 a. antibodies

 b. antigens

 c. interleukins

 d. T cells

78. In order for B cells to be activated, antigens have to bind to the B cell. What kind of antibody does the B cell have that allows for this binding action?

 a. IgA

 b. IgE

 c. IgG

 d. IgM

79. True or False: A fever is a type of immune response that is used to help kill bacteria.

80. Aspirin is used to help reduce pain by inhibiting the production of what chemical?

Chapter 17

81. What is the term used when a person inhales as much as possible and then exhales into a spirometer as much as possible?

 a. residual volume

 b. tidal volume

 c. total lung capacity

 d. vital capacity

82. Henry's law states that _____.

 a. the greater the partial pressure of a gas, the greater the diffusion of that gas

 b. the greater the partial pressure of a gas, the less diffusion of that gas

 c. when partial pressure of a gas is low, the diffusion of that gas is high

 d. when partial pressure of a gas is high, the diffusion of that gas is low

83. Which of the following does NOT describe Boyle's Law?

    a. If the volume of a gas increases, the pressure must also increase in order to maintain equilibrium.

    b. If the volume of a gas decreases, the pressure must increase in order to maintain equilibrium.

    c. If the volume of a gas increases, the pressure must decrease in order to maintain equilibrium.

84. The pneumotaxic centers for respiration are located in the _____.

85. The rhythmic nerve impulses are generated from the _____ _____ to create the normal rhythm of respiration.

Chapter 18

86. What is the name of the enzyme that digests fat?

    a. amylase

    b. lipase

    c. peptidase

    d. trypsin

87. Which of the following does the majority of the digestion of food?

    a. large intestine

    b. liver

    c. small intestine

    d. stomach

88. Which of the following causes the gallbladder to release bile into the small intestine?

    a. cholecystokinin

    b. gastrin

    c. secretin

    d. trypsin

89. Villi or microvilli are structures inside the small intestine that allow _____ of nutrients into the bloodstream.

90. What is the name of the chemical that emulsifies fat to make it easier for lipase to do its job?

Chapter 19

91. What is the name of the catalyst that converts angiotensinogen to angiotensin I?

    a. aldosterone

    b. angiotensin converting enzyme

    c. antidiuretic hormone

    d. renin

92. In order for waste to leave the glomerular capillaries and enter into the capsule, it has to pass through small openings in the capillaries. These small openings are called _____.

    a. cortical openings

    b. fenestrae

    c. podocytes

    d. visceral openings

93. Aldosterone and atrial natriuretic hormone generally have the most effect on which part of the nephron?

    a. glomerular capsule

    b. proximal convoluted tubule

    c. nephron loop

    d. distal convoluted tubule

94. Waste leaves the glomerular capillaries and ultimately ends up in the collecting tubule, eventually making its way to the urinary bladder to be excreted from the body. Put the parts of the nephron in proper sequence as waste travels through the nephron.

95. What are the individual filtering units of the kidneys called?

Chapter 20

96. Successful fertilization of an egg occurs in the _____.

   a. ovaries
   b. uterine tube
   c. uterus
   d. endometrial lining

97. Which of the following is NOT involved in getting the chromosome numbers reduced to a haploid number?

   a. meiosis
   b. mitosis
   c. oogenesis
   d. spermatogenesis

98. What hormone does the corpus luteum release in order to aid in maintaining pregnancy?

   a. follicle stimulating hormone
   b. human chorionic gonadotropin
   c. luteinizing hormone
   d. progesterone

99. What is the name of the enzyme released by sperm in order to decompose the corona radiata of the egg?

100. What hormone is involved in causing the development of a thick endometrial lining for the implantation of a fertilized egg?

Answers: 1. b, **2.** a, **3.** a, **4.** monosaccharide, **5.** enzyme, **6.** b, **7.** b, **8.** a, **9.** osmosis, **10.** metaphase, **11.** d, **12.** a, **13.** b, **14.** endocrine, **15.** dense connective, **16.** c, **17.** d, **18.** b, **19.** arrector pili, **20.** sebaceous, **21.** c, **22.** c, **23.** b, **24.** osteoclasts, **25.** osteons, **26.** b, **27.** c, **28.** c, **29.** False (both sexes have 12 pairs of ribs), **30.** 8, 7, **31.** b, **32.** c, **33.** c, **34.** no movement at all, **35.** diarthrosis, **36.** c, **37.** a, **38.** b, **39.** sarcoplasmic reticulum, **40.** adenosine triphosphate, **41.** a, **42.** a, **43.** a, **44.** ulna (specifically the olecranon process of the ulna), **45.** sternocleidomastoid, **46.** d, **47.** d, **48.** a, **49.** Schwann cells or neurolemmocytes, **50.** neurotransmitters, **51.** b, **52.** b, **53.** c, **54.** meninges, **55.** parasympathetic, **56.** d, **57.** b, **58.** a, **59.** cones, **60.** malleus, **61.** b, **62.** d, **63.** c, **64.** mammary glands and uterus, **65.** erythropoietin, **66.** a, **67.** a, **68.** c, **69.** no, **70.** a trial natriuretic hormone, **71.** d, **72.** a, **73.** b, **74.** lymphatic, **75.** tonsils, **76.** a, **77.** a, **78.** d, **79.** True, **80.** prostaglandin, **81.** d, **82.** a, **83.** a, **84.** pons, **85.** medulla oblongata, **86.** b, **87.** d, **88.** a, **89.** absorption, **90.** bile, **91.** d, **92.** b, **93.** d, **94.** glomerular capsule, proximal convoluted tubule, nephron loop, distal convoluted tubule, **95.** nephrons, **96.** b, **97.** b, **98.** d, **99.** hyaluronidase, **100.** mainly progesterone

THE RESOURCE CENTER

The Resource Center offers the best resources available in print and online to help you study and review the core concepts of anatomy and physiology. You can find additional resources, plus study tips and tools to help test your knowledge, at www.cliffsnotes.com.

Books

This CliffsNotes Quick Review book is one of many great books about anatomy and physiology. For additional resources, check out these publications:

Cohen, Barbara Janson, and Jason James Taylor. *Memmler's Structure and Function of the Human Body,* 9th edition. Lippincott Williams & Wilkins, 2009. Contains modern illustrations, some of which are designed to explain physiological concepts. The information is presented in small segments. There are "Special Interest" boxes that relate the topic to a health concern, as well as summary charts.

Drake, Richard L., A. Wayne Vogl, and Adam W. M. Mitchell. *Gray's Anatomy for Students,* 2nd edition. Churchill Livingstone, 2009. This version of the well-known *Gray's Anatomy* approaches anatomic study by region. The illustrations are designed to assist students in understanding concepts that traditionally have proven difficult for them. There are also helpful sections discussing situations found in clinical or hospital settings.

Shier, David, Jackie Butler, and Ricki Lewis. *Hole's Essentials of Anatomy & Physiology,* 10th edition. McGraw-Hill Science/Engineering/Math, 2008. This reference provides students who have little to no prior experience with anatomy and physiology with an understanding of important concepts. Instead of the traditional approach where one system at a time is studied, the chapters present information in an integrative manner. Includes numerous clinical applications.

Thibodeau, Gary A., and Kevin T. Patton. *Structure & Function of the Body*, 12th edition. Mosby, 2003. This book, written in a very readable style, provides numerous pedagogical aids to student understanding, including outlines, key terms, and pronunciation guides. Includes review questions and study tips.

Venes, Donald, et al. *Taber's Cyclopedic Medical Dictionary,* 21st edition. F. A. Davis Co., 2009. Containing approximately 60,000 definitions and 795 illustrations, *Taber's* is a great dictionary resource for students, as well as an aid in helping students learn the roots of many anatomical terms. There are many helpful appendices, a few of which include immunization schedules, lab tests, standard precautions, weights and measures, and nutrition.

Wolf-Heidegger, edited by Dr. Petra Kopf-Maier. *The Color Atlas of Human Anatomy.* Sterling, 2006. Contains over 650 anatomical drawings. There are numerous color plates illustrating three-dimensional views. The nomenclature within this atlas conforms to the International Anatomic Terminology (Terminolgia Anatomica) of the Federative International Committee on Anatomical Terminology.

Internet

Visit the following Web sites for more information about anatomy and physiology.

anatomy.med.umich.edu/atlas/atlas_index.html shows you views of real brains. The parts of the brain are numbered and provide a key.

homes.bio.psu.edu/people/faculty/strauss/anatomy/skel/skeletal.htm has links to the skeletal system and also to the muscular system. Click on an image to see the labels for the various structures. You can also test yourself by clicking on images that have numbered parts for study purposes; with another click, the correct answers appear. It also has numerous tissue types to click on. Each of the major tissue groups is represented.

http://home.comcast.net/~wnor/thoraxrespiration.htm provides animation for respiratory movements. Be sure to scroll down and click on the arrows.

msjensen.cehd.umn.edu/webanatomy is an excellent site for sample practice quizzes, providing self-tests, timed tests, a multiplayer game, and a Quiz Bowl, all with correct answers. The samples consist of anatomy and also physiology questions.

vcell.ndsu.nodak.edu/animations/home.htm enables you to click on a physiological topic. The main concepts appear in a numbered step-by-step manner. Click on the animation to see the steps "come alive."

www.blackwellpublishing.com/matthews/myosin.html presents muscle contraction animation. Be sure to click on the calcium ion arrows to see how calcium ions are involved in the contraction mechanism.

www.cellsalive.com/mitosis.htm contains excellent animated views of the phases of mitosis and meiosis.

www.getbodysmart.com/ap/muscularsystem/menu/menu.html shows the skeletal muscles of the body. At this site, you are able to remove muscles from the superficial area to see the deeper muscles.

www.montgomerycollege.edu/~wolexik/204 bone_pictures_page. htm is an excellent site for the study of bone structures, with superb images of the various bones of the body in a variety of positions. There are no leader lines or labels provided.

www.youtube.com/watch?v=aQZaNXNroVY&feature=related provides detailed animation of the formation of urine.

www.youtube.com/watch?v=HrMi4GikWwQ contains animation of function of the endocrine system. You can also see hormones entering into the bloodstream and targeting an organ or tissue.

www.youtube.com/watch?v=SDMO4vYkqdg shows animation of impulse activity and the flow of cerebrospinal fluid.

www.youtube.com/watch?v=S3vY5oyXmeI provides animation showing the flow of blood through the heart.

www.youtube.com/watch?v=Z7xKYNz9AS0 enables you to see animation showing the digestion of food.

GLOSSARY

acetylcholine (ACh) a neurotransmitter released by the presynaptic vesicles of axons.

acinar (acini) the smallest portion of a gland. Typically, a secretory cell of the gland.

acne an inflammatory disease of the sebaceous glands.

acrosome a "cap" at the end of the head of a sperm cell, which contains hyaluronidase.

actin a protein substance making up microfilaments commonly found in muscles.

action potential the change in polarity of a neuron resulting in an impulse.

activation energy the amount of energy required to develop a bond between atoms, thus forming a molecule. Defined as the energy that must be overcome in order for a chemical reaction to occur. This could be making or breaking bonds.

adenine one of the nitrogen bases found in DNA.

adenohypophysis the anterior lobe or anterior portion of the pituitary gland.

adenoid a type of tonsil; it is anatomically known as the pharyngeal tonsil.

adenoside diphosphate (ADP) a compound of adenosine with two phosphate groups attached to it. This molecule is used to synthesize ATP.

adenosine triphosphate (ATP) a common molecule that supplies energy for cellular activity.

adenylate cyclase the enzyme that catalyzes the formation of cyclic AMP from ATP.

adrenal cortex the outer layer of the adrenal gland. It secretes aldosterone, cortisol, and androgens.

adventitia the outermost layer of an organ or structure.

aerobic refers to a chemical reaction or pathway requiring the presence of oxygen.

afferent transporting toward some central organ. Afferent nerves transmit impulses toward the CNS. Afferent blood vessels transport blood toward the heart. Afferent lymph vessels transport lymph fluid to the lymph nodes.

afterload the measure of the pressure that is required by the ventricles to get the semilunar valves to open.

agglutination a type of reaction that causes the red blood cells to clump to each other.

agglutinogens a special type of antigen located on the surface of red blood cells.

agranulocytes white blood cells that do not have granules in their cytosol (lymphocytes and monocytes).

allele one of two or more different genes found on corresponding positions on paired chromatids.

all-or-none law the concept used to describe the fact that once an impulse is developed, it will go all the way to its destination. The all-or-none law is the principle that the strength by which a nerve or muscle fiber responds to a stimulus is not dependent on the strength of the stimulus. If the stimulus is any strength above threshold, the nerve or muscle fiber will give a complete response or otherwise no response at all.

alveolar ducts the final branches of the bronchial tree of the respiratory system.

alveolar macrophage cells that remove debris and microorganisms from the alveolar walls of the respiratory system.

amphiarthrosis slightly movable joint.

ampulla the expanded base of the semicircular canals; also an expanded area of a tube.

amygdala a group of neurons associated with the limbic system within the temporal lobe.

amylases a group of enzymes that digest carbohydrates.

amylose another word for starch.

anabolism the synthesis of new products.

anaerobic refers to chemical reactions or pathways that do not require the presence of oxygen. Glycolysis (during metabolism), for instance, is a series of reactions that are anaerobic because oxygen is not a part of the reaction.

anastomoses the area where two or more vessels merge.

anatomy the study of the structure and relationship between body parts.

androgen-binding protein (ABP) a chemical that binds to testosterone (T), dihydrotestosterone (DHT), and 17-beta-estradiol making them less lipophilic.

antagonist something that reacts opposite to something else; for example, a muscle.

antagonistic hormones hormones that acts the opposite of other hormones.

antibiotics a chemical designed to kill bacteria.

antibodies a complex glycoprotein produced by B lymphocytes that binds to specific antigens.

antibody-mediated response an immunity resulting from activated B lymphocytes.

antigens protein markers on the surface of cells marking them as "self" or "nonself."

antigen-presenting cell (APC) a cell (typically a macrophage) that breaks down antigens and presents the fragments on the surface so other cells can respond to it.

antiporters enzymes that move ions in opposite directions.

apneustic area an area located in the pons that stimulates inspiration.

apoenzyme the protein portion of an enzyme.

aponeurosis a flat, broad extension of tissue that functions the same as tendons. Typically found connecting one muscle to another.

aquaporins water channels located in the cell membrane used to assist movement of water into and out of a cell.

arteriosclerosis narrowing of the arteries.

ascending tract a bundle of nerves in the spinal cord that transmits information to the brain.

association neurons neurons in the CNS that transmit impulses from one sensory neuron to another.

astrocytes a type of glial cell that maintains an ion balance within the neuron.

atom the smallest component of an element.

autonomic nervous system (ANS) the portion of the nervous system that we cannot control. These nerves respond automatically to environmental factors.

autorhythmic cells special cells that produce their own impulse and act as the heart's natural pacemaker.

avascular referring to the lack of blood vessels.

axolemma the plasma membrane of a neuron.

axon the elongated portion of a neuron that transmits impulses.

axon hillock the region where the axon emerges from the neuron body. Action potentials begin here.

axoplasm the cytoplasm of a neuron.

B cells special lymphocytes that produce antibodies in response to foreign antigens that have entered the body.

baroreceptors sensory receptors that monitor blood pressure in the carotid artery and aortic arch.

basal bodies structures at the base of flagella and cilia that appear to organize the development of the microtubules that make up flagella and cilia.

basal ganglia pockets of gray matter located within the brain that help control movement.

bicarbonate ion antiporter a structure of the cell membrane that transports bicarbonate ions across a membrane.

bile a substance produced by the liver and stored in the gallbladder. Bile emulsifies fat in the digestive system.

bilirubin a byproduct produced by the breakdown of old, worn-out red blood cells.

bisphosphoglycerate a byproduct of glycolysis in the red blood cells.

blood-brain barrier capillaries of the brain that are designed in such a manner as to prevent the passage of toxic materials, but allow the passage of oxygen and nutrients, etc. Due to the structure of these capillaries, many antibiotics cannot cross the barrier.

Bohr effect a condition when hydrogen ions bind to hemoglobin, resulting in an increase of carbon dioxide and/or hydrogen ions in blood, which causes decreased affinity of hemoglobin for oxygen.

bolus a chewed mass of food that is to be swallowed.

Boyle's Law a law stating that at a constant temperature, the volume of a gas varies inversely with the pressure.

brainstem the part of the brain that consists of the midbrain, pons, and medulla oblongata.

Brunner's (duodenal) glands glands found in the submucosa of the duodenum that secrete material to neutralize the gastric acid coming from the stomach.

buccal phase the phase of swallowing where the tongue pushes the bolus of food toward the pharynx.

bulbourethral gland a small reproductive gland that secretes semen into the urethra.

cancer uncontrolled cell growth and reproduction.

capsular hydrostatic pressure the pressure that develops as water collects in the glomerular capsule of the nephrons.

carbaminohemoglobin (HbCO$_2$) the amino acid portion of hemoglobin bound to carbon dioxide.

carbohydrates organic molecules having the formula CH$_2$O.

carbonic anhydrase an enzyme that assists in the union of water and carbon dioxide to form carbonic acid. It can also be used to reverse the reaction.

carboxyl group a functional group of organic molecules with the formula –COOH.

carcinogens agents that may cause cancer.

cardiac output (CO) the volume of blood pumped out of the heart per minute.

catabolism a type of metabolism where substances are broken down.

catalyst a substance that lowers the activation energy of a chemical reaction so the reaction is accelerated.

cell the smallest unit of life, consisting of organelles.

cell-mediated an immune response involving T cells that respond to MHC markers on pathogens or tumor cells.

cellular respiration a process in which the cells produce ATP.

centrioles structures that produce protein spindle fibers involved in mitosis.

centromere the area where paired chromatids are attached together.

centrosomes a pair of centrioles that is located outside the nuclear envelope.

cephalic phase stimuli that influence the digestive activities, originating in the head area (smell, taste, and thoughts).

chemoreceptors sensory receptors that respond to dissolved chemicals.

chemotaxis the movement of cells (such as leukocites) to a specific area due to the release of a host of chemicals from damaged cells or other leukocytes.

chiasmata sites where genetic material is exchanged between nonsister homologous chromatids.

chondroitin a type of proteoglycan that is part of ground substance in connective tissue.

chromatids the two identical longitudinal halves of a chromosome.

chromatin the threadlike material of DNA. The chromatin will condense to form chromosomes.

chromosomes rod-shaped bodies that consist of DNA, located within the nucleus.

chyle the lymph and its contents that are absorbed into the lacteals of the villi of the small intestine.

chylomicrons fat-soluble vitamins that are mixed with cholesterol and protein.

chyme digested food within the stomach, which is then passed to the small intestine.

coagulation blood clotting.

codon a group of three adjacent nucleotides on the messenger RNA molecule.

coenzyme an organic molecule that assists enzymes in speeding up chemical reactions.

cofactor a nonprotein molecule that assists enzymes in speeding up chemical reactions.

colony-stimulating factor hormones produced by white blood cells, thus initiating the process of leukopoiesis to make more white blood cells.

complement protein a group of proteins that assist in the immune system.

complementary strand a single strand of the new, replicated DNA.

complete tetanus the frequency of stimuli is of such a manner as to cause muscle fibers to contract in a smooth and swift manner to reach a plateau rapidly, making it so that individual twitches are no longer distinguishable.

connective tissue tissue that consists of scattered cells with a matrix located between the cells.

corpus callosum a tract of nerve fibers that "connects" the two cerebral hemispheres.

corpus luteum the remains of a ruptured follicle after the ovulation of an egg that secretes estrogen and progesterone.

costimulation a process that requires two signals to activate a cellular reaction. Typically, one is from an antigen and the other is from an immune cell such as a T cell.

countercurrent multiplier system the movement of fluid in opposite directions through tubes that are side by side.

cranial vault the top, sides, front, and back of the skull.

creatinine a byproduct of the breakdown of creatine phosphate during muscle contraction.

crypts of Lieberkühn the spaces between the villi within the small intestine.

cytokine chemicals produced by a variety of cells that stimulate other cells to be responsive in the immune system.

Dalton's law a law stating that the sum of the partial pressures of each gas in a mixture is equal to the total pressure of all the gases.

dartos muscle a muscle in the superficial fascia of the scrotum.

defensins antibiotic proteins released from the granules of neutrophils in an effort to kill bacteria.

deglutition the process of swallowing.

denticulate ligament ligaments that hold the spinal cord in position within the vertebral canal.

dentin calcified tissue that makes up most of the tooth.

deoxygenated without or reduced oxygen, such as deoxygenated blood (veins).

deoxyribonucleic acid (DNA) molecules that make up chromosomes and store genetic information.

depolarize the opening of the gated channels that allows sodium ions to enter into the neuron.

dermatomes any area on the body that receives nerve stimuli. Many times, these areas are parallel to each other.

desmosomes a cellular junction made of glycoprotein that provides attachment between adjacent cells.

detrusor muscle a layer of muscle that makes up part of the wall of the urinary bladder.

dialysis the movement of solutes across a cell membrane.

diaphysis the shaft area of a long bone.

diarthrosis a freely moveable joint.

diastole relaxation of the heart during a heartbeat.

diploid a cell containing two copies of each chromosome; the cell, therefore, contains a full set of the original number of chromosomes.

distal a point that is farthest from the point of attachment of a limb, or a point that is farthest from a specific point of reference.

edema an increase in body fluids in a localized area resulting in swelling.

effector a muscle, gland, or organ that receives an impulse.

efferent a vessel or nerve that transports fluid or impulses away from a specified area.

eicosanoids lipids that are synthesized from fatty acid chains.

electrocardiogram a recording of the electrical cardiac cycle events.

electrolytes substances that will become ions when exposed to water.

emulsification the process of breaking up large globules of fat and turning them into smaller globules of fat.

end-diastolic volume (EDV) the volume of blood contained within each ventricle at the end of diastole.

endogenous antigen antigens that are synthesized by cancerous cells (for example).

end-systolic volume (ESV) the volume of blood left in the ventricles after ventricular contraction.

enteric nerves nerves that innervate the GI tract.

enteroendocrine cells cells of the digestive system (enteric) that release hormones (endocrine) into the bloodstream.

epiphyses the ends of the long bones.

EPSP (excitatory postsynaptic potential) a depolarized neural membrane.

erythropoiesis the process of red blood cell formation in bone.

esophageal hiatus an opening in the diaphragm that allows passage of the esophagus to the stomach.

esophageal phase the phase in which food passes through the esophagus via peristaltic action.

exogenous antigen antigens that have been engulfed from bacteria or viruses (external source).

exons the coding regions of genes with DNA.

expiratory reserve volume (ERV) the amount of air that can be exhaled after a normal exhalation.

exteroceptors special sense organs that provide information about the external environment.

facet a smooth surface (sometimes curved) where two bones come together to form a joint.

facilitated diffusion the diffusion of solutes through protein channels across a cell membrane.

fascia a fibrous membrane that covers, supports, or separates one muscle from another. Superficial fascia is found underlying the skin. Deep fascia separates one muscle from another.

fenestrae openings in capillaries (especially in the capillaries that make up the glomerular capillaries).

flagellum the tail of sperm cells.

fontanel a membrane-covered area within the spaces created by a fetal skull. Typically called the "baby's soft spot."

foramen a hole in a bone that allows passage of blood vessels, nerves, or both.

forebrain the anterior portion of the brain of the embryo.

fossa a depression in a bone that is involved in articulation.

fovea a pitlike depression in a bone (smaller than a fossa).

Frank-Starling law a law stating that the greater the heart chambers stretch, the greater the contraction will be.

functional residual capacity (FRC) the amount of air remaining in the lungs after normal expiration.

gamete a sex cell (egg or sperm).

gap junctions membrane proteins that bind the cell memebranes of adjacent cells together but form a channel between the two cells.

gastric phase the process of digestion in the stomach.

gated channels channels that act like gates to allow ions to pass through the cell membrane.

glomerular filtration rate (GFR) the rate at which filtrate accumulates in the glomerular capsule (approximately 180 L/day).

glomerular hydrostatic pressure blood pressure in the glomerulus.

glomerular osmotic pressure the pressure in the glomerulus as a result of the movement of water and solutes out of the glomerular capillaries.

glycogenesis the conversion of glucose to glycogen for storage.

glycogenolysis the conversion of glycogen to glucose.

glycolysis the chemical reactions involved in converting glucose to pyruvic acid within the cell.

graded potential a decremental change in the resting potential of a neuron in response to a stimulus.

granulocytes a class of leukocytes that consist of granules in their cytoplasm. These granules consist of a variety of chemicals.

haploid a cell containing only one copy of each chromosome, thereby giving the cell one half the original number of chromosomes.

heart rate (HR) The number of ventricular contractions of the heart per minute.

helicase an enzyme that "unwinds" the DNA when the DNA is preparing to duplicate.

helicotrema an area where the scala vestibuli and scala tympani join to form a continuous tube within the cochlea.

hematopoiesis the formation of blood (same as hemopoiesis).

hemocytoblasts stem cells within the bone marrow that differentiate into the various blood cells.

hemoglobin the iron-containing molecule of red blood cells that carries oxygen from the lungs to the tissues.

hemostasis the process of blood clotting.

Henry's law a law stating that the greater the partial pressure of a gas, the greater the diffusion of the gas into a liquid.

(Hering-Breur) inflation reflex a response to the stretching of the walls of the bronchi and bronchioles when air enters the lungs. The response is to allow expiration to begin.

hindbrain the posterior portion of the brain of the embryo. It can be differentiated into the cerebellum, pons, and medulla oblongata.

homeostasis the state of dynamic equilibrium of the internal environment. This state of equilibrium is maintained by a variety of feedback mechanisms.

homologous pair of chromosomes a pair of chromosomes, each having the same characteristics as the other.

hormones chemical messengers that are released from endocrine cells or glands into the bloodstream and travel to target cells.

hydrogen bond a weak bond that forms between the oxygen of one water molecule and the hydrogen of a different water molecule.

hydrophilic refers to a molecule or a part of a molecule that is attracted to water and is dissolved by water (*hydro* = water and *philia* = love; hence "water-lover").

hydrophobic refers to a molecule or a part of a molecule that repels water and is not dissolved by water (*hydro* = water and *phobia* = fear; hence "water-fearer").

hyperpolarization a change in a cell's membrane potential that makes it more negative than the resting membrane potential.

hypoxia having an inadequate amount of oxygen.

immunocompetent the ability of the body's immune system to respond to pathogens and damaged tissue. When T cells become activated, they are said to be "immunocompetent."

immunoglobulins special classes of antibodies (IgA, IgD, IgE, IgG, and IgM).

incomplete tetanus the frequency of stimuli is of such to cause the muscle fiber contractions to reach a plateau and not contract any stronger. See *complete tetanus.*

inhibitory postsynaptic potential inhibiting an action potential usually due to hyperpolarization. A synaptic potential that decreases the chance that a future action potential will occur in a postsynaptic neuron.

insertion the point of attachment of a muscle that is the moveable end of the joint.

inspiratory capacity (IC) the maximum amount of air that can be inhaled with one breath.

inspiratory reserve volume (IRV) the amount of air that can be inhaled after a normal inhalation.

interferons (IFNs) a substance secreted by cells after the invasion of a virus. This substance activates nearby cells to begin to battle the virus.

interleukin a chemical that helps leukocytes communicate with each other, thus activating each other to help defend the body.

interneurons neurons in the CNS that transmit impulses from a sensory neuron to a motor neuron, or to other interneurons. Also called an association neuron.

interoceptors receptors that respond to stimuli within the visceral organs. Also called visceroceptors.

introns noncoding sequences within DNA.

juxtaglomerular apparatus (JGA) cells within the kidneys (near the glomerular capsule of the nephrons) that detect and respond to low blood pressure.

kinetochores specialized regions within the centromeres where spindle fibers attach during cell reproduction.

kinins a group of chemicals released in response to cell injury. The response activates other cells to begin the repair process.

Kupffer cells phagocytic cells within the liver.

lacteals lymph capillaries that are found in the villi of the small intestine. Lacteals absorb lipids.

lacunae a depression area in bone or cartilage where osteocytes or chondrocytes reside.

leukocytes white blood cells.

leukopoiesis the formation of leukocytes.

lung compliance the measure of the ability of the lungs and thoracic cavity to expand.

lymph a type of fluid tissue, which involves cells of the immune system.

major histocompatibility complex (MHC) molecules found on the surface of cell membranes used to identify cells as "self" or "nonself."

megakaryoblasts a large stem cell that will fragment and become platelets.

meiosis a type of cell reproduction that results in each new cell having a haploid number of chromosomes.

meninges the three layers of membranes covering the brain and spinal cord.

menstrual cycle the cycle of events where the immature eggs mature, develop, and eventually ovulate. The unfertilized egg will not implant and the thickened endometrial lining will shed.

menstrual phase the phase where the thickened endometrial lining is shed from the body.

messenger RNA (mRNA) an RNA molecule that transports genetic information for making a protein from the DNA to the ribosomes.

metabolism chemical reactions involved in the formation of new substances.

midbrain the uppermost part of the brainstem.

mitosis the activity of cell reproduction that occurs within the nuclear region of the cell.

monosynaptic reflex arc an involuntary reflex that involves only one synaptic area.

motor neuron a neuron that transmits impulses away from the CNS.

mucosa associated lymphoid tissue (MALT) bundles of lymphatic cells found in the mucous linings of the intestinal, respiratory, reproductive, and urinary tracts.

mutagens radiation or chemicals that cause mutations.

mutation a change in a gene, due to a DNA error, which is potentially capable of being passed on to the offspring.

negative feedback a situation where a sensing mechanism detects a change in conditions, causing an action in the body to correct the change.

negative inotropic factors hormones or drugs that decrease the contractility of the heart.

nephrons specialized tubes within the kidneys.

net filtration pressure (NFP) the pressure that forces filtrate out of the capillaries and into the glomerular capsule, for example.

neurotransmitter a chemical released from the axon vesicles that enters into the synaptic region. These chemicals will then act to stimulate or inhibit

the postsynaptic membrane of the next neuron in sequence.

oogenesis the formation of eggs via meiosis.

optic disc the area where blood vessels enter the eye and the optic nerve exits the eye. This area is void of rods or cones, so no "visual activity" occurs here.

organelles specialized structures within a cell that perform specific functions.

origin the point of attachment of a muscle that is the stationary end of the joint.

osmosis the diffusion of water molecules across a cell membrane.

osteon the main unit of compact bone in which osteocytes are found.

oxyhemoglobin (HbO$_2$) the combined form of oxygen and hemoglobin.

paracrines hormones that affect only nearby cells.

passive immunity immunity that is obtained by transferring antibodies from a person who has an illness, or the transfer of antibodies to an infant via the placenta or breast milk.

pathogen a disease-causing organism.

peristalsis a muscular action that moves food along the digestive tract.

Peyer's patches a cluster of lymphatic nodules that are found in the lining of the ileum of the small intestine.

pH the measure of hydrogen ions (hydronium ions) in a designated volume of solution.

phagocytes leukocytes that wander around in the tissues and "eat" pathogens.

phagocytosis a process where some cells have the ability to ingest foreign matter, such as bacteria. These cells can do this by enveloping their cell membrane around the bacterium.

phosphorylation the process of adding a phosphate to another molecule. Usually this phosphate comes from ATP.

physiology the science of the function of living organisms and its components. It includes all of the chemical and physical processes involved.

platelets fragments of megakaryocytes that are involved in blood clotting.

pinocytosis the movement of fluids into the cell via the enclosing of the cell membrane, thus forming vesicles at the membrane surface.

polymorphonuclear a nucleus that consists of several lobes connected by thin strands of material. This gives the one nucleus the appearance of being many multishaped nuclei.

positive feedback a situation where a change has occurred but the response to the change exaggerates the change even more.

positive inotropic factor hormones or drugs that increase the contractility of the heart.

proprioceptors special sense organs that make us aware of the position of the bones, joints, and muscles.

receptor-mediated endocytosis the invagination of the cell membrane, t hus forming a vacuole around specific molecules, which are then brought into the cell.

renin-angiotensinogen a hormonal mechanism of the kidneys that regulates blood volume and thus blood pressure.

residual volume (RV) the volume of air still within the lungs after maximum exhalation.

resting potential an unstimulated polarized neuron.

reticular activating system (RAS) a component of the brainstem, which is involved in maintaining alertness.

Rh blood group a complex group of antigens on the surface of red blood cells in addition to the A and B antigens.

ribonucleic acid (RNA) A nucleic acid that is involved in the control of protein synthesis. There are three major types. See *messenger RNA, ribosomal RNA,* and *transfer RNA.*

ribosomal RNA (rRNA) an RNA molecule comprised of the ribosomes involved in making protein.

sarcolemma the membrane of muscle cells.

sarcomere the functional unit of muscle cells that extends from one Z line to the next Z line.

seminiferous tubules tubules found within the testes. The cells of these tubules produce sperm cells upon exposure to the follicle stimulating hormone.

serum the liquid material remaining after blood-clotting proteins have been removed from the plasma.

solute the material that is being dissolved in solution.

solvent the material doing the dissolving.

somatic nervous system (SNS) a system of nerve fibers that run from the CNS to the skeletal muscles.

spermatogenesis the process of forming sperm cells via meiosis.

spermatogonia stem cells that will become sperm cells.

steroid a group of chemicals that consists of a ring of carbon within its molecular structure.

stroke volume (SV) the volume of blood ejected by the ventricles with a single contraction.

substrate the substance being acted upon. For example, an enzyme acts on chemical A to form chemical B. Chemical A is the substrate.

surfactant a surface-active agent that lowers the surface tension of the fluid lining the alveoli.

synapse the gap between one neuron and the next neuron.

synarthrosis an immovable joint.

synergists muscles that assist the prime mover muscle in its action.

synovial joint freely moveable joints consisting of a joint cavity filled with synovial fluid.

systole the contraction of the atria and ventricles.

T cells lymphocytes that are exposed to thymosin from the thymus gland, thus turning into special immune cells called T cells.

T wave a wave of activity on an ECG recording illustrating the repolarization of the ventricles.

target cell/organ the cell or organ that is affected by a hormone.

telodendria branches at the end of axons.

threshold level the amount of stimuli required to open the gated channels of a neuron. The amount of stimuli required to initiate an impulse.

thrombocytes a name formerly used for platelets. The term implies cells (cytes). Technically these are not cells; they are cell fragments. Therefore, the term has been changed to platelets.

thrombopoiesis the formation of thrombocytes (platelets).

tidal volume (TV) the amount of air passively exhaled (approximately 500 mL of air).

tight junctions a type of junction where the cell membranes of adjacent cells are bound together via interlocking proteins.

tonsils lymphoid tissue in the mucus lining of the pharynx region and also at the base of the tongue. The palatine tonsil consists of two lymphoid tissues on either side of the uvula. The lingual tonsil is lymphoid tissue located at the base of the tongue. The pharyngeal tonsil commonly referred to as the adenoids is located on the roof of the nasopharynx region.

total lung capacity (TLC) the maximum amount of air the lungs can hold (approximately 6,000 mL).

transcription a process where the DNA is used as a template to create RNA.

transfer RNA an RNA molecule that transfers amino acids to the ribosomes to be assembled into protein.

translation a process where the ribosomes use the information encoded in the mRNA to assemble the proteins.

trigone the triangular base of the urinary bladder where the two ureters enter and the urethra exits the urinary bladder.

urobilinogen the product produced after bacteria have broken down bilirubin.

vaccines substances that activate memory B cells in the immune system.

vascularized having blood vessels.

vasoconstriction the constriction of blood vessels.

vasoconstrictor a substance that causes the constriction of blood vessels.

vasodilation the dilation of blood vessels.

veins blood vessels that transport blood toward the heart.

vital capacity (VC) the amount of air that can be exhaled after a maximal inhalation (approximately 4800 mL).

Wormian bone bones that form within sutures (sutural bone).

zygote a fertilized egg.

Index

Notes

Notes

Master the basics—fast!